Beyond Minerals: Raw Reliance's Eco-Cost

Naina

Copyright © 2024 by Naina

Table of Content

CHAPTER 1: Introduction and literature review

1.1. Introduction

Nowadays, the international market's demand for raw materials as a primary consequence of globalization and industrialization and digitalization are growing much faster compared to the past decades. As a good example, copper's consumption, as one of the most significant materials in the world, has increased from almost ten million tonnes in 1980 to 23 million tonnes in 2015 (Figure 1-1).

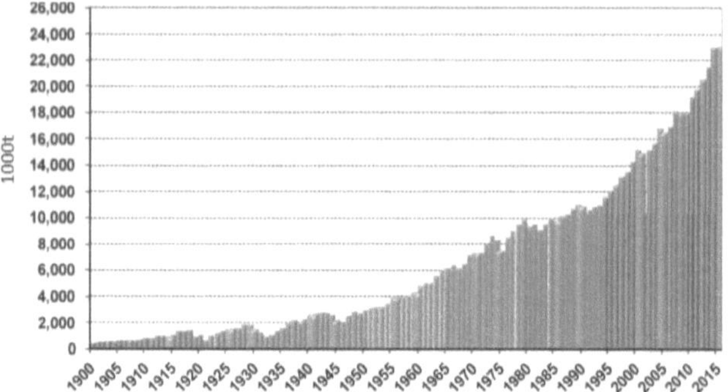

Figure 1-1. World consumption of refined copper in the last century [1]

To respond to the increasing demand, the mining industry worldwide has to increase its production, leading to a considerable decrease in natural resources due to the massive amount of extraction of minerals. Additionally, it caused a reduction in the ore grade of the extracted ore. In copper, whereas the world production has increased (Figure 1-2), the world average copper grades show a decreasing trend in the last decades and forecasting for the coming years (Figure 1-3).

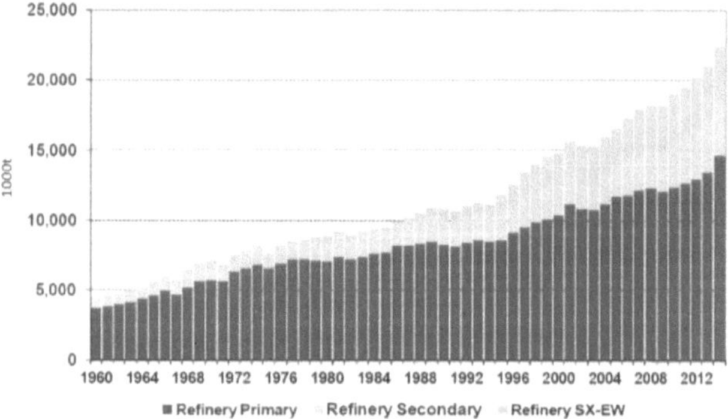

Figure 1-2. World production of refined copper in the last decades [1]

This situation can even be worse when the prices are oscillating, and in many cases, they are decreasing (Figure 1-4).

All these facts force mine managers to plan in a way that not only can fulfill the increasing demand of the world but making their profit. For that, many efforts were carried out to reduce the cost of mining. Although reducing costs through various policies can help mine managers to overcome such problems, it is not the only challenging issue that they are confronting. In recent years, many environmental concerns regarding mining projects arise. For instance, there are numerous oppositions against coal mining in Germany [2] mainly due to its environmental impacts and social problems. It is another side of mining that forces miners to consider environmental and social issues in their planning. Nowadays, mining is highly demanding for a "sustainable design", which not only technical and economic issues must be taken into account, but also environmental, safety, and social concerns need to be emphasized. It makes the miners' work more difficult than before due to considering different aspects of their design and its implication.

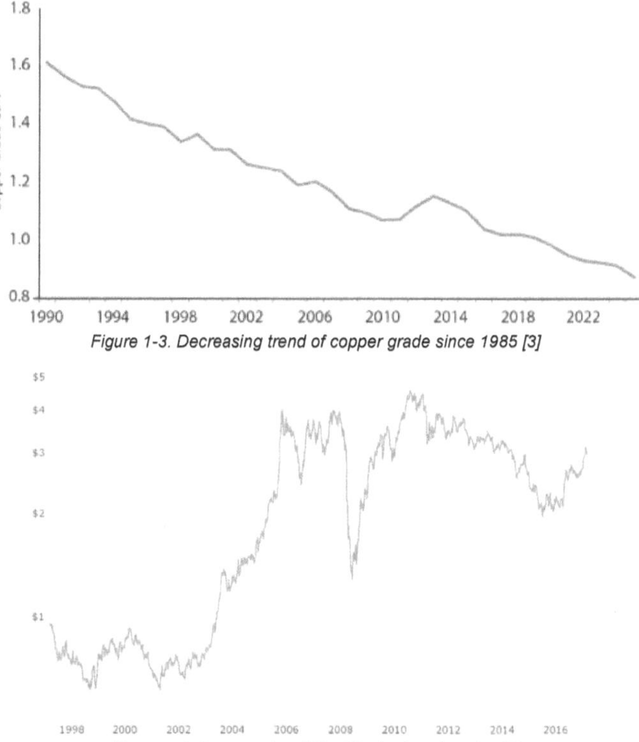

Figure 1-3. Decreasing trend of copper grade since 1985 [3]

Figure 1-4. Copper price ($/lbs.) in the last decades [4]

The transportation system in a mining project is one of the most significant parts of the mine design. On the one hand, it needs a high amount of operating and capital cost and, on the other hand, affects the environmental and safety issues in the mining area. Although the Truck-Shovel system, known as the conventional transportation system, is commonly used in mining projects, there are other types of transportation systems called In-Pit Crushing and Conveying (IPCC) systems. Generally accepted, these systems present lower operating costs and environmental impacts because of working by electricity in contrast with trucks. A fact about trucks is that they impose higher operating costs and more environmental consequences predominantly due to burning fossil fuels. Furthermore, the safety level in the IPCC systems in case of accidents, injuries, and fatalities considerably increases by eliminating or reducing the number of trucks.

Accordingly, it is necessary to rethink these systems for utilization in mines to achieve a desirable design.

1.2. Literature review[4]

The Truck-Shovel system, known as the conventional loading and hauling system, is widely used in open-pit mines. However, by increasing the mine's life and increasing the technical, economic, environmental, safety, and social importance, these systems' application needs to be re-evaluated. The situations that can convince mine planners to look for an alternative for the conventional Truck-Shovel system are:

- New technical aspects in mining projects include the higher production demand, deepening the mine, increasing the stripping ratio, lower grade, unplanned delays and unavailability of the Truck-Shovel systems, etc.
- Changing in the economic conditions in operating costs in the last decades, such as increasing fuel price, spare parts, etc.
- Environmental restrictions that need to be obeyed during the mine activities include reducing emissions, dust, noise, etc.
- Safety and social circumstances that can affect the mining project, such as lowering the quantity of labor force, considering public health, safety, road traffic, training the employees, etc.

The IPCC systems, as the alternatives for the Truck-Shovel system, can resolve the deficits as mentioned above of this system to a high extend predominantly through the following features:

- The capability of moving a high volume of material (ore or waste) because of the continuous system of hauling as well as higher availability.
- Lower operating costs in an IPCC system mainly because of lower electricity prices.
- Lower production of emissions, dust, and noise in comparison with the Truck-Shovel system.
- Providing a safer working environment because of not using trucks as a moving object inside the mining area and lower the need for the labor force.

It exceeds more than half a century that the mining industry started applying IPCC systems in mining projects. The first utilization of the IPCC systems goes back to the 1960s in a limestone quarry in Höver, Germany [5, 6] (as cited in [7]). In recent decades, the willingness to use this type of crushing and haulage systems will be increased (Figure 1-5). Figure 1-6 shows the distribution of the different kinds of IPCC systems since 1970 around the world [7]. Ritter provided all these IPCC systems, such as mine's name, type of feeder, crusher, IPCC, capacity, etc., in his work [7].

Some researchers mentioned the IPCC systems as the future of mining. They believe that the future economic situation in mining projects will encourage mine managers to

---

[4] This subsection (1.2) was published as a conference paper cited as:

Abbaspour, H., Drebenstedt, C. (2019). IPCC systems as a bulk material handling method in mines: a review regarding the technical, economic, environmental, safety and social factors. *Proceedings of the VIII International Symposium of Young Researchers TRANSPORT PROBLEMS 2019*, 785-796.

rethink the conventional Truck-Shovel system and find a way to substitute it with the IPCC systems [8, 9]. In a couple of relevant works, Tutton et al. studied the capability of using different types of IPCC systems in hard rock mines [10]. Zimmermann evaluated the possibility of the application of the IPCC systems in quarries [11].

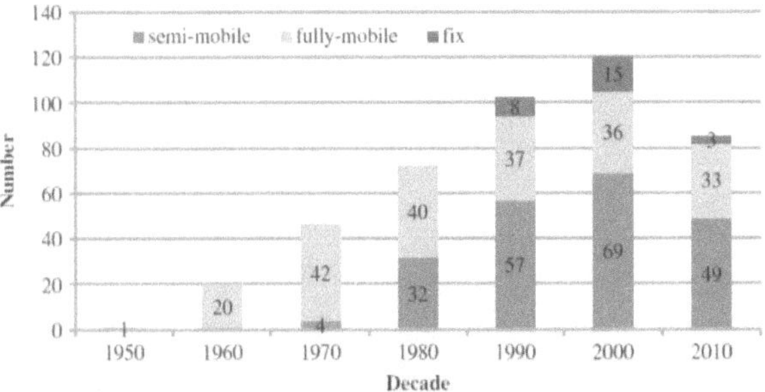

Figure 1-5. IPCC installation in the last decades [12]

The growth in the application of IPCC systems led to more investigations in these systems. Each of these studies paid attention to different aspects of the IPCC systems. Generally, they can be categorized into five elements of the technical, economic, environmental, safety, and social issues, which are thoroughly discussed in the following subsections.

1.2.1. Technical aspect of the IPCC systems

The selection of IPCC systems is one of the first technical challenges for a mining project. Atchison and Morrison presented a collection of factors that must be considered while selecting an IPCC system. The most critical factors they mentioned are "productivity of the systems, interactions with the drill/blast and bench operation sequence, the ease of relocation, flexibility to changes in the reserve, scalability, and compatibility with other elements of the system" [13].

Using the IPCC systems can change the mine design concept, especially in the final pit. Johnson investigated that pit shell design can be different when a mining project uses an IPCC system instead of the conventional Truck-Shovel system. He believes that using the IPCC system can affect the "unit cost of a deposit," and it should be considered in the whole process of mining design [14]. Figure 1-7 clearly shows the difference between the mine's final shell based on Truck-Shovel or IPCC systems [14].

Hay et al. introduced a new method to determine the ultimate pit limit (UPL) of an open-pit mine benefiting from a Semi-Mobile In-Pit Crushing and Conveying (SMIPCC) system [15]. They showed that not only the design of the UPL would be different in comparison with the Truck-Shovel system, but also offers a higher net present value (NPV) for their case study. In a similar study, Hey introduced a new method for determining the UPL of open-pit mines by considering the Fully Mobile In-Pit Crushing and Conveying (FMIPCC) system as the crushing and transportation in the project [16]. Dean et al. presented a method for selecting the FMIPCC systems in deep open-pit mines [17]. They also specified a planning method that is a combination of the Truck-Shovel system to excavate a "box cut" in one phase and subsequently, using the FMIPCC system for "parallel and radial pushbacks" in other phases [17]. Figure 1-8 shows the concept of their planning method for a deep mine.

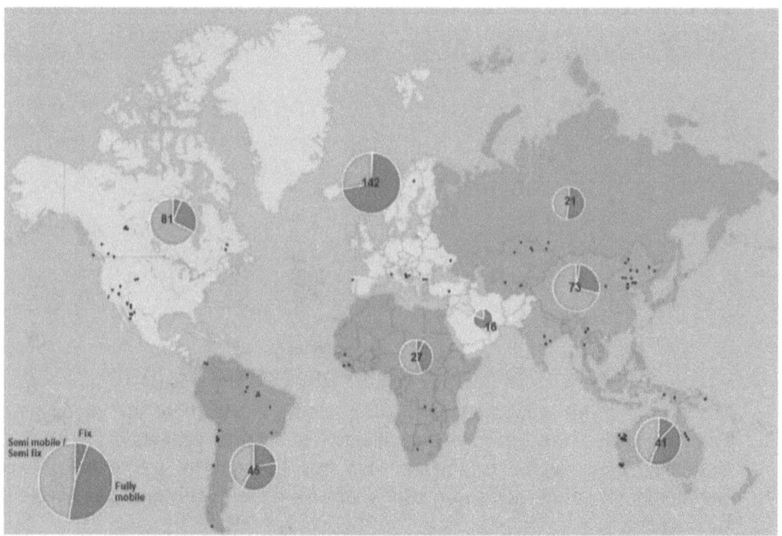

Figure 1-6. Distribution of the IPCC systems around the world since 1970 [7]

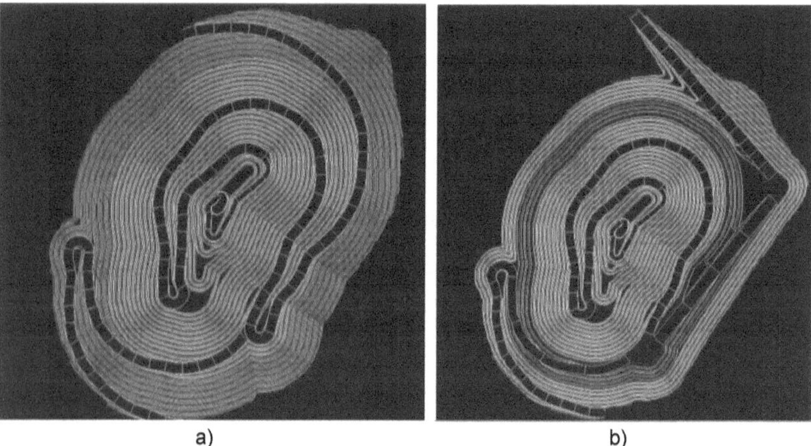

a)   b)

Figure 1-7. The final pit shell design a) in the Truck-Shovel system and b) in the IPCC system [14]

Nehring et al. compared three scenarios of mine planning by using the Truck-Shovel, SMIPCC, and FMIPCC systems [18]. They showed that using each of these transportation systems can change the mine planning approach, which in their case study, the FMIPCC represented a higher lifetime and NPV for the project followed by the SMIPCC and Truck-Shovel system (Figure 1-9) [18].

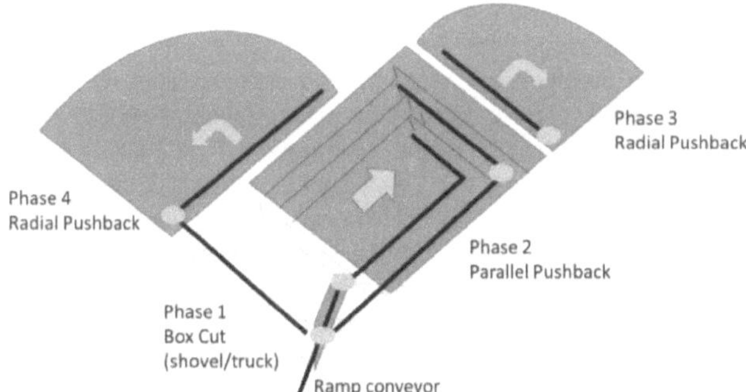

Figure 1-8. Combination of the Truck-Shovel system (Phase 1) and FMIPCC system (Phases 2-4) [17]

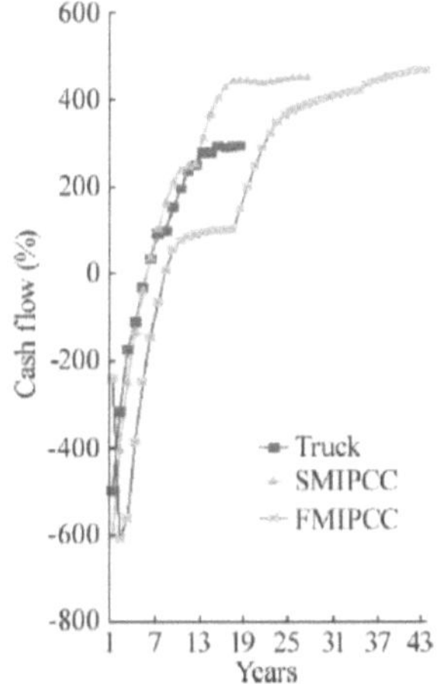

Figure 1-9. The cash flow and mine's life in the Truck-Shovel, SMIPCC, and FMIPCC systems [18]

The IPCC systems' proper location and relocation plan are the other technical concerns that attracted researchers to work on them. Sturgul discussed the IPCC system's optimum location in a mine by implementing trucks' cycling time and simulation of the loading and hauling cycle [19]. Changzhi presented work in Dexin pit copper mine about the Truck-Shovel system's transition point to the IPCC system [20]. Additionally, he mentioned the IPCC system's relocation time and changes in these points by changing the price [20]. Konak et al. studied finding the optimum location of an IPCC system in an aggregate mine. They presented an optimum location for the IPCC system to minimize the haulage distance based on the geometry of the mine and access requirement [21]. They defined all the possible locations for installing the IPCC system and selected the best option based on a "trial and error" process [21]. Rahmanpour et al. tried to determine an IPCC system's optimum location through a "single hub location problem" [22]. The objective of this problem was to minimize the haulage costs and environmental considerations. They investigated about 17 factors that can affect the optimum location of the IPCC system. However, they stated that the "mine plan, haulage distance to faces and mine facilities" are dominating factors determining IPCC location [22]. Paricheh et al. presented a heuristic method for calculating the optimum location and time for using the IPCC systems. They divided this problem into two different levels: firstly, the optimum location and relocation of the IPCC (OL) through the mine life based on the dynamic facility location problem and secondly, the optimum time (OT) for adding the IPCC systems in mine projects based on the maximization of NPV [23]. In the former, they defined the objective function as the minimization of the haulage costs and, in the latter, maximizing NPV regarding the Truck-Shovel and IPCC systems production [23]. In another work, they studied the effect of uncertainties in the production and operating cost on the IPCC systems' optimum location [24]. Abbaspour et al. examined the relocation plan of a semi-mobile in-pit crushing and conveying system by the transportation problem [25]. They defined this problem as a matrix of transportation problem, which includes the possible sources (year and level of production), destinations (the locations of the IPCC system) as well as the supply (production related to the sources) and demand (the capacity of IPCC system) related to them respectively. Builes defined a mixed-integer programming model to determine a SMIPCC system's optimum location for an iron mine [26].

Some other works focused on the flexibility and availability of the IPCC systems while operating in mines. Morriss assessed the availability of IPCC systems in mining operations and its effect in the other parts of mining such as the shovel, conveyors, spreader, etc. [27]. He defined the IPCC system as a "chain", in which their parts have interactions. He evaluated these interactions in terms of availability and utilization of every single piece that can favorably affect the IPCC system's annual output [27]. McCarthy pointed out the flexibility issue in the IPCC systems. He provided some guidelines in the different IPCC systems to have better flexibility while operating [28]. He believed that despite the lower cost of IPCC systems than trucks, loss of "flexibility is a real" issue [28]. Dzakpata et al. made a comparison of the Truck-Shovel and IPCC systems based on the "operating efficiency" in a mining project. They introduced the "utilized time, operating time, and valuable operating time" parameters to define the operating efficiency [29]. They stated that while trucks benefit from higher flexibility, conveyor belts show a higher performance in the parameters mentioned earlier.

Dos Santos et al. examined the capability of using different IPCC systems with a high-angle conveyor belt [30]. In Yugoslavian copper mine, three distinct phases of mining with varying types of haulage systems were designed [30]. At phase 2 (1980 to 1987) and phase 3 (after 1988), a fixed crusher for moving waste and a portable crusher with a shiftable conveyor for transporting ore were designed [30]. Paricheh and Osanloo evaluated different options of exiting the conveyor belt in an IPCC system from an open-pit mine [31].

1.2.2. Economic aspect of the IPCC systems

The economic concern of the IPCC systems is one of the most challenging factors, which many researchers focused on this issue. This criterion is one of the most important reasons for using these systems by reducing operating costs. For instance, in 1980, an IPCC system in Bingham Canyon Mine was employed, which reduced cost in this project [32]. Radlowski compared all truck haulage and IPCC systems in the capital and operating costs in an open-pit mine [33]. He showed that utilizing an IPCC system instead of all trucks would result in a 30% lower total cost [33]. He used simulation for building his model to evaluate the truck haulage in terms of operating time and fuel consumption [33]. Some other researchers also worked on this theme and tried to investigate the IPCC systems' cost-saving potential. Martin and Utley showed that cost saving could be in a range of 26%-50% [34, 35] (as cited in [33]). Some other researchers assessed the ability of the IPCC systems as a continuous haulage system. Terezopoulos studied the use of high-speed conveyor belts (mainly the side-slewable belt conveyors) with a combination of the IPCC system [36]. He stated that this combination could change the profitability of the project by reducing the operating costs. He mentioned the lower power consumption as another advantage of using these systems. He also made a comparison between continuous and non-continuous haulage systems in mines. Lieberwirth and Almeida et al. compared using conveyor belts in the IPCC systems instead of trucks for transporting ore and highlighted the advantages of using these systems [37, 38]. Nunes et al. compared using the Truck-Shovel and SMIPCC system in operating costs in a Brazilian copper-gold mine. They showed that using the SMIPCC system will impose a lower operating cost to the project [39].

1.2.3. Environmental aspect of the IPCC systems

The environmentally friendly feature of the IPCC systems attracted a few researchers to focus on this issue. Norgate and Haque studied the greenhouse gas emission caused by the IPCC systems [40]. They estimated the reduction of greenhouse gas generation by using the IPCC systems. A life cycle assessment study showed that using these systems results between 4%-22% fewer greenhouse gas emissions than the Truck-Shovel system, in which the former is related to the electricity generated by coal, and the latter corresponding to the electricity generated by natural gas [40]. They also mentioned that this emission is highly dependent on the transportation distances and the annual capacity of the system. In relatively similar research, Raaz and Mentges studied the difference between the trucks and IPCC systems using energy and CO_2 emission [41]. Awuah-Offei et al. examined the environmental benefits of utilizing conveyor belts against trucks for transferring ore in a gold open-pit mine through the life cycle assessment [42]. They showed that using trucks would result in 300% more potential for acid rain. They also demonstrated that using natural gas instead of coal for producing electricity for conveyor belts could reduce the environmental impact of using them. De Almeida et al. evaluated the equivalent CO_2 emission and waste generation (tire and belts) of the conveyor belt and trucks, which conveyor belt produced more wastes but less CO_2 emission [38].

1.2.4. Safety and social aspects of the IPCC systems

The general opinion regarding the safety of IPCC systems introduces them as the safe systems in contrast with trucks. However, the number of studies that thoroughly investigated the safety of IPCC systems are few. Most of them discuss the conveyor belts as a part of these systems. Kecojevic et al. developed a risk assessment method to identify the risk of using conveyor belts in injuries and fatality accidents [43]. They investigated the data between 1995 to 2006 of injuries and fatalities related to the conveyors in the USA's mines. They benefited from a preliminary hazard assessment (PHA) method for quantifying the risk. Hill investigated the improvement in safety and causes of injuries and fatalities in using conveyor belts through conveyor belts' safety analysis in the last three decades [44].

1.3. Problem statement

The IPCC systems are not a new technology in the mining industry; however, they are not the priority as the transportation system. It could be because of the following reasons:

1. The high demand for capital cost compared with the Truck-Shovel system: capital cost is one of the most dominant factors in deciding for system selection. If it is taken into account that buying and installing an IPCC system will cost more than 150 million dollars at once, probably this option would be spontaneously ignored.
2. The lower flexibility, in contrast with the Truck-Shovel system: the high degree of flexibility, is one of the favorable benefits of trucks. This factor provides this possibility for mine planners to assess various options of production planning freely. Additionally, any out-of-service truck can be substituted by another truck to prevent interruption in the mine's production plan. However, in the IPCC systems, flexibility is lower, and planning must be performed based on particular conditions. For instance, the IPCC system must be located at a point that cannot affect the extraction process. This condition needs to be accounted for in planning, which will reduce the options for design.
3. Demand for a high amount of electricity: while the IPCC systems use electricity for operating, it is essential to predict their electricity demand. In some cases, it would even be necessary to construct a local power station.

At first glance, it could be convincing not to use the IPCC systems; however, this is not the whole story, and there are still other factors that must be considered. For this, many investigations were carried out to study these systems from different viewpoints. Although many studies about the comparison of the Truck-Shovel and IPCC systems are performed in the case of technical and economic issues [8, 13, 14, 17, 20, 27, 29], investigation on the environmental, safety, and social impacts of using these systems are few [40, 44]. Furthermore, most of these works are operated in a static situation, whereas it is very significant to have a dynamic evaluation of the conditions. Finally, there is no tool that can compare these systems in all the technical, economic, environmental, safety, and social issues.

1.4. Project's objective

A literature review on the IPCC systems in different points of view, including the technical, economic, environmental, safety and social, was conducted in the previous section. In some aspects, more attention is paid to the IPCC systems (e.g., the technical and economic factors). In contrast, it was not considerable on the others (e.g., the environmental, safety, and social aspects). On the other hand, in most of them, just one type of IPCC system was considered and compared with the Truck-Shovel system, while there are differences in various types of IPCC systems. Besides, not considering all these aspects (i.e., technical, economic, environmental, safety, and social) as an integrated system for comparing different transportation systems (Truck-Shovel and IPCC systems), will not be resulted in the best outcome. Accordingly, this project aims to develop a methodology to select and compare IPCC and Truck-Shovel transportation system alternatives using a holistic approach that considers technical, economic, environmental, safety, and social impacts. Such impacts may be measured in non-financial terms. This goal will be fulfilled by modeling the transportation system in system dynamics modeling in Vensim© DSS software (version 6.2). Furthermore, the software will be designed and developed to access and run the model for public use. As the second step, the output of the system dynamics model in Vensim will be imported as the input in another software developed by the author in Java programming language, which implements the final decision of selecting these transportation systems based on the Analytic Hierarchy Process (AHP).

To sum up, this project is going to fulfill the following major objectives:

1- Defining a dynamic model of the transportation system in open-pit mines by considering the Truck-Shovel and IPCC systems.
2- Describing the technical, economic, environmental, safety, and social indexes for these transportation systems in the dynamic model.
3- Comparing these systems regarding their outcomes in the indexes mentioned above.
4- Software development, as a result of building dynamics model, for presenting to the public.
5- Software development based on AHP, using Vensim software's output as input, to perform the transportation system selection through the mine's life.

1.5. book structure

The state of the art of two concepts of the IPCC systems and system dynamics modeling, which are mainly focused on this project, will be presented in Chapter 2. Chapter 3 is assigned to developing the system dynamics model for different indexes (technical, economic, safety, environmental, and social) and explaining how it is designed. All these indexes, including their related factors, will be discussed in this chapter. The AHP will be shortly described in Chapter 4, and the process of evaluating the indexes regarding the AHP will be defined. In Chapter 5, a case study for examining the system dynamics model and AHP will be presented. Finally, in Chapter 6, a summary of the project, and a list of recommendations for future works related to this project will be provided.

CHAPTER 2: State of the art of the IPCC systems and system dynamics modeling

2.1. State of the art of the IPCC systems

2.1.1. IPCC definition

The IPCC systems can be described as the semi-continuous to continuous haulage systems, constituted from the crusher station, conveyor belt, and discharge system. In contrast with a standard crusher station, which is located in a place outside the pit, this system can be set inside. These systems can be fed by shovels, dozers, draglines, trucks, or a combination of this equipment [45], which depends on the type of IPCC system. The first IPCC system was introduced in 1956 in Germany [45] and used as an alternative for the conventional transportation system in the surface mining operations [46].

These systems are typically categorized based on their crushing system. These are including Fixed In-Pit Crushing and Conveying (FIPCC) system, Semi Fixed In-Pit Crushing and Conveying (SFIPCC) system, Semi-Mobile In-Pit Crushing and Conveying (SMIPCC) system, and Fully Mobile In-Pit Crushing and Conveying (FMIPCC) system [47]. The general classification of the Truck-Shovel and IPCC systems are represented in Figure 2-1, which shows the main parts of these systems.

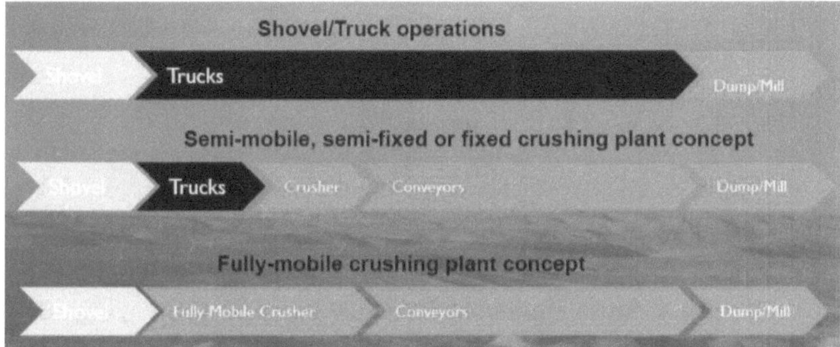

Figure 2-1. IPCC systems and their operating parts [48]

2.1.2. Fixed In-Pit Crushing and Conveying (FIPCC) system

In the FIPCC system, the location of the crusher is fixed along with the mine's life. Commonly, the position of this type is near the pit rim or inside, which is not affected by mining operations [45]. This type is typically divided into two forms: in-ground and rim-mounted crushing plant. In the former, "the crusher is installed in a concrete structure below grade and located external to the pit and never moved, but the latter is a part of or attached to the bench wing wall and usually installed for 15 or more years" [47]. Figure 2-2 shows a rim-mounted FIPCC system.

2.1.3. Semi Fixed In-Pit Crushing and Conveying (SFIPCC) system

The SFIPCC system is located in a strategic junction point in the pit and is mostly fed by the mining trucks. Its relocation needs disassembly of the entire crusher station into several parts or multiple modules [45]. "Some degree of disassembly is required to move the structure. The planned frequency of moves for a semifixed crusher is no less than five to 10 years" [47]. Figure 2-3 shows a schematic cross-section and plan of a SFIPCC system.

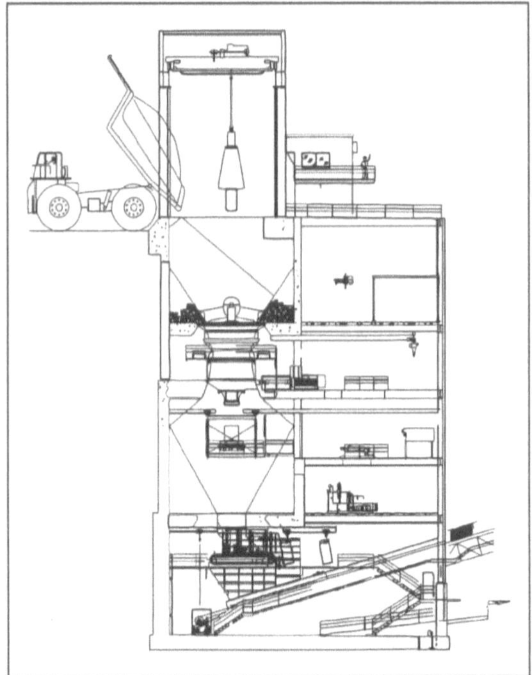

Figure 2-2. A schematic view of a rim-mounted FIPCC system [47]

2.1.4. Semi Mobile In-Pit Crushing and Conveying (SMIPCC) system

The SMIPCC system does not have an integrated transportation system and is commonly located at the operational level. It is an all-steel structure and possible to be fed through trucks or loaders from different loading points. The planned frequency of moves for a movable crusher is between three to five years [47]. Figure 2-4 shows a SMIPCC system in an open-pit mine.

2.1.5. Fully Mobile In-Pit Crushing and Conveying (FMIPCC) system

The FMIPCC system can continuously change its location and benefits from an integrated transportation mechanism [7]. Despite the other IPCC systems, this system is fed directly by a loading machine (shovel, dozer, etc.), and trucks do not use it in the loading and haulage process. Figure 2-5 depicts a typical FMIPCC with a combination of bridge and shiftable conveyor belt.

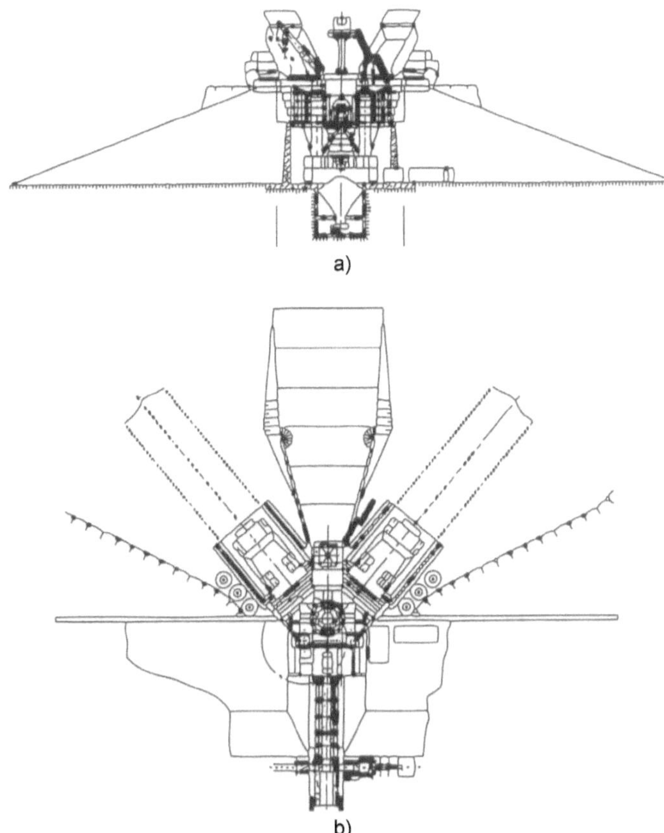

a)

b)

Figure 2-3. a) The section and b) the plan view of a SFIPCC system [47]

Figure 2-4- A SMIPCC system in an open-pit mine [48]

Figure 2-5. A FMIPCC system (including bridge fully mobile and shiftable conveyor belts) [48]

Some of the pros and cons of these IPCC systems are described in Table 2-1.

Table 2-1. Advantages and disadvantages of the IPCC systems [48]

System	Advantages	Disadvantages
FIPCC	• Traditional plants can simply be modified in order to be used as in-pit crushers. • Reducing maintenance costs because of no need for an apron feeder • High crushing chamber capacity • Reducing capital costs mainly because of the limited capability of mobility • Reduced maintenance costs because of a better crushing in the upper portion of the chamber and decreased localized abrasive wear • Higher capacity and finer product size due to the weight of the ore column above the crusher	• Concrete design cannot be relocated • Structural steel cannot usually be moved. If it is not the case, a considerable sub-structure is needed to support the plant for relocation • Overall height is more significant because of the higher dump point bench level • Large retaining wall • Long bench strikes and width in min plan
SFIPCC	• Traditional plants can be simply modified in order to be used as in-pit crushers • Reducing maintenance costs due to no need for an apron feeder • High crushing chamber capacity • Reducing capital costs mainly due to the limited capability of mobility	• Only the crusher and part or all of the hopper are mounted on a steel base • The balance of the station is civil construction. Greater overall height is due to the higher dump point bench level. • Long bench strikes and width in min plan

	Advantages	Disadvantages
	• Increased long-term flexibility due to the limited capability of moving, which allows for future modifications. • Reduced maintenance costs due to a better crushing in the upper portion of the chamber and decreased localized abrasive wear • Higher capacity and finer product size due to the weight of the ore column above the crusher	
SMIPCC	• Traditional plant configuration • Low bench height for dumping ore • Reduced truck queue time due to the surge pocket • Improved control of oversize material fed to the crusher • Reduced crusher downtime due to the bridging of large lumps • High crushing chamber throughput • Reduced capital costs due to the limited degree of mobility • Increased long-term flexibility due to the ability to move the complete station intact, • Reduced maintenance costs due to deletion of the apron feeder, the greater amount of crushing in the upper portion of the chamber, and decreased localized abrasive wear when compared to indirect feed designs • Greater capacity and finer product size due to the weight of the ore column	• Large and heavy structure requiring large transporters for moving • Greater overall height due to the higher dump point bench level, which requires extensive bench-retaining walls • Long bench strikes and width in min plan
FMIPCC	• Elimination of truck transport • Reduced number of personnel • Avoidance of high truck maintenance costs • Reduction of mine traffic • Increase in the overall safety	• Increased total capital costs • Increased maintenance costs associated with adding an apron feeder • Increased maintenance costs associated with the crusher from using an apron feeder • Long bench strikes and width in min plan

2.1.6. Conveying system

In the IPCC systems, different types of conveying systems based on the project's needs are used. Generally, there are five different types of conveying systems [7], which are:

- Fully mobile
- Portable
- Shiftable
- Semi fixed
- Fixed

2.1.6.1.   Fully mobile conveyor belt

This conveyor belt is capable of changing its location through the mining operation continuously. This type is generally used in combination with the FMIPCC systems [7]. The main fully mobile conveyor belts are belt wagon, bridge conveyors, and horizontal conveyors (Figure 2-6).

a)

b)

c)

Figure 2-6. Different types of fully mobile conveyor belts: a) belt wagon (TAKRAF) [49] b) bridge [50] c) horizontal (TNT) [7]

2.1.6.2. Portable conveyor belt

Like the fully mobile conveyor belt, this conveyor belt also can continuously move while extraction but using tires instead of crawlers (Figure 2-7).

Figure 2-7. A portable conveyor belt in a limestone mine (Metso) [51]

2.1.6.3. Shiftable conveyor belt

This conveyor belt is constituted from several shiftable conveyor modules, each four to six meters long. These modules are installed on a steel sleeper, which is concocted to the steel rail. This steel rail allows the whole system can be shifted by pipe laying dozers without dismantling the modules [7]. This type of conveyor belt is usually used with the SMIPCC or FIMIPCC systems. Figure 2-8 shows this type of conveyor belt in the operating face.

Figure 2-8. Shiftable conveyor belt installed at the operating face [52]

2.1.6.4.   Semi fixed conveyor belt

As it is clear from its name, it is occasionally relocated whenever it is needed. It is constituted from portable modules of length four to six meters, which are installed on the concrete sleepers. A picture of a semi-fixed conveyor belt is shown in Figure 2-9.

2.1.6.5.   Fixed conveyor belt

This type of conveyor belt should not be relocated along with the mine's life (Figure 2-10). It is typically installed outside the pit and works with a fixed crusher station [7].

Figure 2-9. Semi fixed conveyor belt [7]

Figure 2-10. Fixed conveyor belt [53]

2.1.7. Discharge system

The last point in an IPCC system is the discharge system, in which the material is unloaded from the conveying system. Based on the material, different types of discharge systems are introduced. The most important of them are [7]:

- Spreader, which operates at the dumpsite and used for the overburden or waste (Figure 2-11 a)
- Stacker, which distributes the low-grade ore at the heap leach pad or ore at the stockyards. (Figure 2-11 b)

- Stacker-Reclaimer, which is utilized for "unloading material onto storage piles and reclaiming when required" [7] (Figure 2-11 c).

a)

b)

c)

Figure 2-11. Discharge systems a) spreader [54] b) stacker [7] c) stacker-reclaimer [54]

2.2. State of the art of system dynamics modeling

2.2.1. General Description

System dynamics is a method to describe dynamically, model, simulate, and analyze the complex issues and/or the systems in terms of the processes, information, organizational boundaries, and strategies [55]. It is a methodology that can handle many features that a standard analysis hardly can or even cannot, e.g., feedback analysis, time-delay, nonlinearity, etc. [56]. This method was introduced in the 1950s to help managers and decision-makers to improve their understanding of their surrounding processes. Nowadays, system dynamics is extensively using in all public and private sectors to policy analysis and design.

System dynamics modeling is related to the dynamic behavior of the systems, i.e., in contrast with static analysis, in which the time does not have any role, it is considered throughout the analysis. In system dynamics modeling, the modeler tries to recognize the patterns of behavior being represented by essential variables in the system and then build a model that can simulate the patterns. When a model has this ability, it can be used as a laboratory for testing policies to alter a system's behavior in desired ways [57].

A system can consist of different aspects like physical, economic, social, technological, etc. Such a system is highly complex [56]. For instance, "agricultural production systems with climate change in an agricultural system consists of physical, biological, social, technological, environmental, economic, and political components and their interactions" [56].

Generally, there are two different types of system dynamics models. First, open systems (Figure 2-12 a), in which the outputs do not affect their inputs in contrast with the closed systems (Figure 2-12 b), where the outputs can control and modify the inputs.

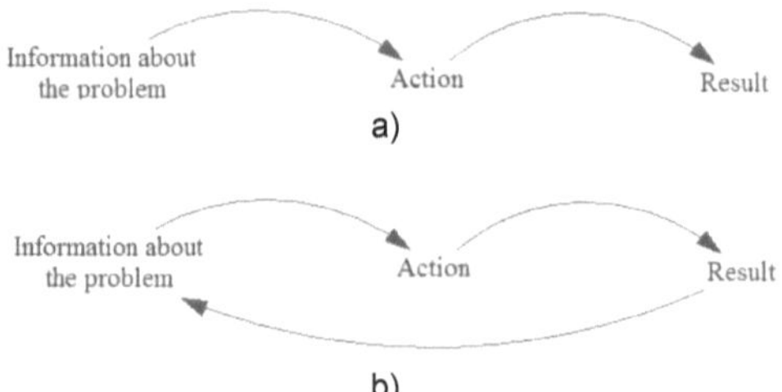

Figure 2-12. A schematic view of the concept of a) an open system and b) a closed system (adopted from [56])

System thinking is another important concept in system dynamics, which is a method for evaluating a system's dynamic behavior rather than its single components [56]. By considering each component as an individual part, a false outcome of a dynamic system could happen. Accordingly, "systems thinking should consider all the interacting components that influence the dynamics of the complex system" [56].

Every system dynamics model consists of constants, auxiliaries, stocks, flows, and feedbacks that will be explained in the following sub-sections.

2.2.2. Constants and auxiliaries

All of the parameters that form a system dynamics model are divided into two groups of constants or auxiliaries. Constants are permanently fixed in the whole time of the system processing. Auxiliaries are defined as an equation among different constants or other auxiliaries, which might change during system processing.

2.2.3. Stocks and Flows

In system dynamics models, it is essential to distinguish between two types of variables: stocks and flows.

Stocks represent the results of accumulations over time. Their values are the "level" of accumulation, which are also called "states" as they collectively represent the state of the system at time t [58]. Some examples of stock could be cash balance, production, etc.

Flows directly flow into and out of the stocks, thus changing their levels. They represent the "rate of change" of the stocks [58]. Income and expenses, production, and sale rate are examples of flows. The unit of the rates must be defined as $unit/time$. This feature fulfills the possibility of accumulation for the stocks during the time.

The graphical representation of the stock-flow diagram is as Figure 2-13.

Figure 2-13. A schematic view of a simple stock and flow diagram

The general relationship of any stock-flow diagram can be described as the following equation:

$$\frac{d(Stock)}{dt} = \text{Inflow (t)} - \text{Outflow (t)}$$

Equation 2-1

2.2.4. Feedback

Although stocks and flows are both necessary and sufficient for generating dynamic behavior, they are not the only building blocks of dynamic systems. More precisely, the stocks and flows in the real world systems are part of the feedback loops, which are often joined together by the nonlinear couplings that cause counterintuitive behavior [57]. Figure 2-14 shows a simple stock-flow system with feedback that connects the stock to the inflow, meaning that the stock's level as a useful parameter can change the amount of inflow based on the received feedback.

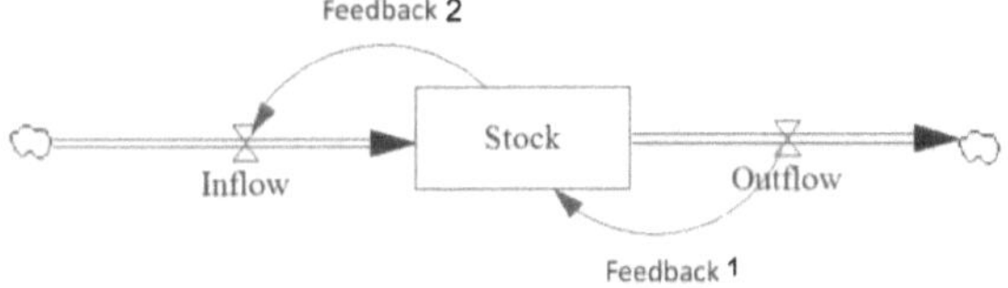

Figure 2-14- Feedback in a stock-flow diagram

$$\frac{d(Stock)}{dt} = \text{Inflow (t)} - \text{Outflow (t)} + \text{Feedback 1 (t)} - \text{Feedback 2 (t)}$$

Equation 2-2

2.2.5. Positive and Negative Loops

"Closed systems are controlled by two types of feedback loops: positive and negative loops. Positive loops portray self-reinforcing processes where an action creates a result that generates more of the action, and hence more of the result. Anything that can be described as a vicious or virtuous circle can be classified as a positive feedback process. Generally speaking, positive feedback processes destabilize systems and cause them to "run away" from their current position. Thus, they are responsible for the growth or decline of systems, although they may occasionally work to stabilize them. In Figure 2-15a), the loop is positive and defines a self-reinforcing process. For example, if a shock were to suddenly raise Variable A, Variable B would fall (i.e., move in the opposite direction as Variable A), Variable C would fall (i.e., move in the same direction as Variable B), Variable D would rise (i.e., move in the opposite direction as Variable C), and Variable A would rise even further (i.e., move in the same direction as Variable D)" [57].

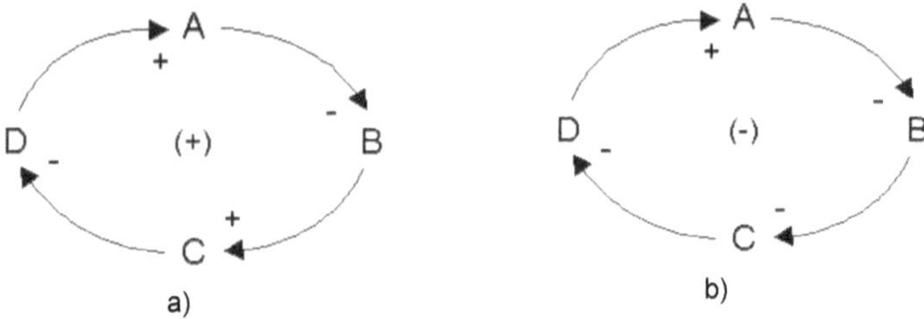

Figure 2-15. Feedback loops a) positive b) negative [57]

"Negative feedback loops describe goal-seeking processes that generate actions aimed at moving a system toward or keeping a system at the desired state. In general, negative feedback processes stabilize systems, although they may occasionally destabilize them by causing them to oscillate. Figure 2-15b) presents a generic causal loop diagram of a negative feedback loop structure. If an external shock were to make Variable A fall, Variable B would rise (i.e., move in the opposite direction as Variable A), Variable C would fall (i.e., move in the opposite direction as Variable B), Variable D would rise (i.e., move in the opposite directions Variable C), and Variable A would rise (i.e., move in the same direction as Variable D). The rise in Variable A after the shock propagates around the loop acts to stabilize the system, i.e., move it back towards its state prior to the shock. The shock is thus counteracted by the system's response" [57].

2.2.6. Cause and use trees

The concept "tree" defines that from which components they are caused and in which ones are used. These trees could be constructed from a couple up to hundreds of parameters. For instance, Figure 2-16a shows the loaded truck power, which is caused by some parameters such as the loaded truck rimpull, loaded truck speed, and the number of trucks. In contrast, Figure 2-16b shows that it is used in forming some other parameters such as the fuel consumption (loaded truck), power consumption per meter of transferring material, and technical index.

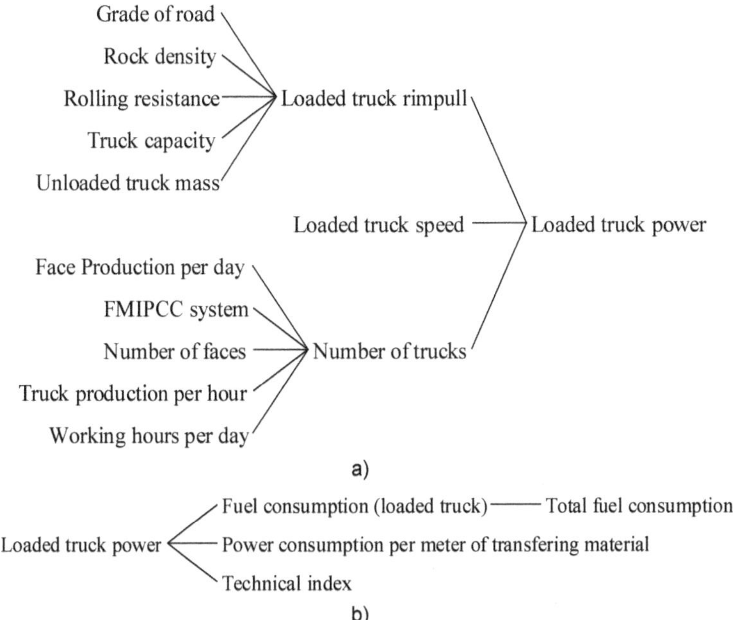

Figure 2-16. a) Cause tree and b) use trees of the "loaded truck power"

CHAPTER 3: System dynamics model construction

3.1. Steps of system dynamics modeling

Designing a system dynamics model depends on any individual modeler and the steps he/she considers. In addition, there is not a specific method that confines modelers in constructing a system dynamics model. However, the most general and accepted steps by building a system dynamics model based on a system thinking method are [59]:

I.   Conceptualization
  - Purpose of the model
  - Model boundary
  - Reference modes
  - Basic mechanisms
II. Formulation
  - Levels and rates equations
  - Estimate and select the value of parameters
III. Testing
  - Simulate the model
  - Testing model assumptions
  - Test model behavior and sensitivity analysis
IV. Implementation
  - Testing model in different policies
  - Presenting outputs in an accessible form

Except for the first step (conceptualization), which is introduced in the next four sections, other steps (formulation, testing, and implementation) are not strictly separated. They are discussed throughout the construction of the stock-flow diagram, putting equation, feedback loops, simulation, and sensitivity analysis in the next sections.

3.2. Purpose of the model

The first and most crucial step in constructing any system dynamics model is defining that model's purpose. In this regard, two main aspects need to be covered: focusing on the problem and defining the model's audience [59]. The general purpose of the model is:

"Introducing technical, economic, environmental, safety and social indexes (TEcESaS indexes) for the transportation systems (Truck-Shovel and IPCCs) in open-pit mines"

However, the specific purpose of the model is:

"Comparison of the different transportation systems based on the technical, economic, environmental, safety and social indexes (TEcESaS indexes) to select one of them throughout the mine's life"

Since the model's audience are the scientific and industrial groups, which cover a wide range of students, professors, and engineers, it is tried not only the model be readable and traceable for them but also covers their requirements.

3.3. Model boundary

From drilling and blasting to processing, the mining operation is the connected parts, which changing in each of them can affect the others. The whole mining can be considered an integrated system, in which each part cannot be separated from the other parts. However, in this study, the focus is only on the transportation and primary crushing systems that include loading (shovels), hauling (trucks and conveyor belts), and primary crushing (IPCC) systems.

3.4. Reference modes

Reference modes show the behavior of the critical variables in the model through time. They can be described in the verbal or graphical views, while the former is interpreted through time by using words and sentences, and the latter converts this interpretation into the graphs. The reference mode can be constructed from previous real data or analysis and expected behavior of the phenomenon called historical and hypothetical reference modes [59]. Typically, these reference modes are used as guidelines before and after model construction to assess the model's results. Whenever the model's final result is closed to the reference mode, it can be claimed that the model is appropriately built; otherwise, it would need to be rebuilt or modified.

3.5. Basic mechanisms

Basic mechanisms are generally constituted from different parameters and their relations that can finally produce the reference mode's behavior. Basic mechanisms are typically instituted by the stock-flow diagram (SFD) or the causal-loop diagram (CLD). While many modelers prefer the former, others are in favor of the latter. It is up to the modeler's decision that which one should be taken. However, the relation and connection among parameters are essential. Figure 3-1 shows examples of the SFD and the CLD, respectively, for a pocket money model [60].

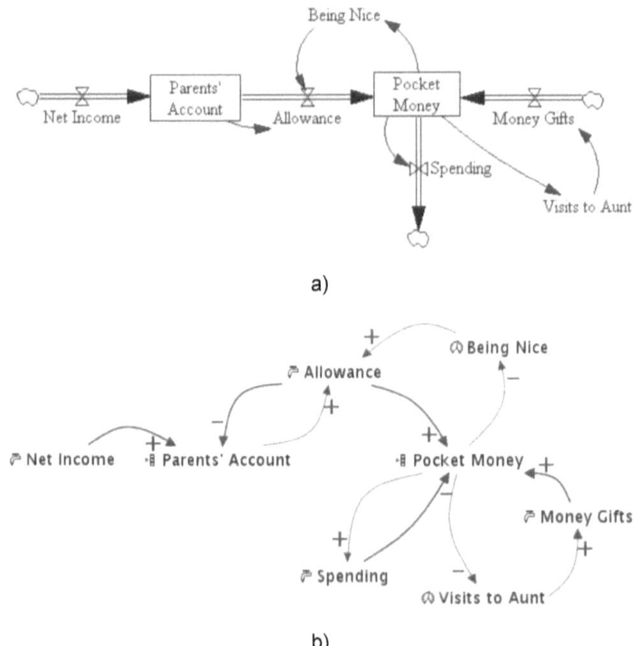

Figure 3-1. An example of a) SFD and b) CLD for pocket money [60]

3.6. General note on constructing the system dynamics model

It is necessary first to choose the constants, auxiliaries to construct the SFD or CLD of any system dynamics model, flows, and stocks. These collections and their relations constitute the SFD or CLD.

Since the purpose of the model is to compare the different indexes in transportation systems (Truck-Shovel and IPCCs), they are based on a comparative attitude to describe a proper comparison between the different transportation system alternatives. Although it is possible to consider as much as constants, auxiliaries, stocks and flows in the model, it depends on the way of the system thinking and the ability to find relations among them, which is considered a challenging issue for modelers. Even though numerous items can be selected as the influencing factors on the indexes, more relevant, measurable, and vital factors are considered in the model.

Figure 3-2 depicts the flowchart of defining the index equation of different factors $i = \{1,2,...,n\}$. Generally, the first step is to determine whether an equation or statistical data can be put in the model; otherwise, the next factor would be tried. The index equation can be defined by providing the cause tree of any factor and distinguishing its positive or negative impacts on the index (reinforcing or balancing). Additionally, this process for any other sub-factor resulted from the cause tree of the factor i, would be tested.

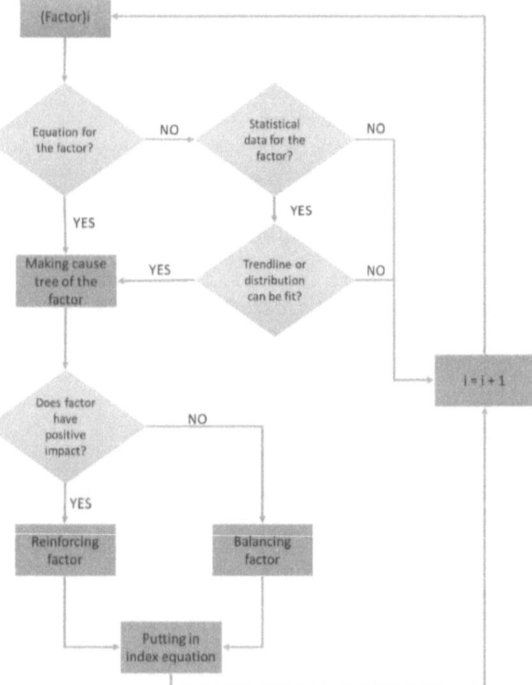

Figure 3-2. Flowchart of defining factors, causes trees and index equation

3.7. Technical index

For defining the technical index for the transportation systems, three factors of system availability, system utilization, and power consumption were introduced. These factors are constituted from trucks, conveyor belts, and IPCC systems (Figure 3-3).

System availability and utilization are defined as the overall availability and utilization that a transportation system offers. For instance, the availability and utilization in a SMIPCC system is a combination of the availability and utilization of the shovels, trucks, and SMIPCC and conveyor belts as a system. Since the power consumption in the IPCC systems is determined by the type of the crusher, it would be the same for the different types of IPCC systems in a mine with a specific rock property. Accordingly, trucks and conveyor belts' power consumption were considered as the two most critical comparative factors. The general sketch of the system dynamics model constructed for the technical index in Vensim is as Figure 3-4. Each part will be explained in detail in the next sections.

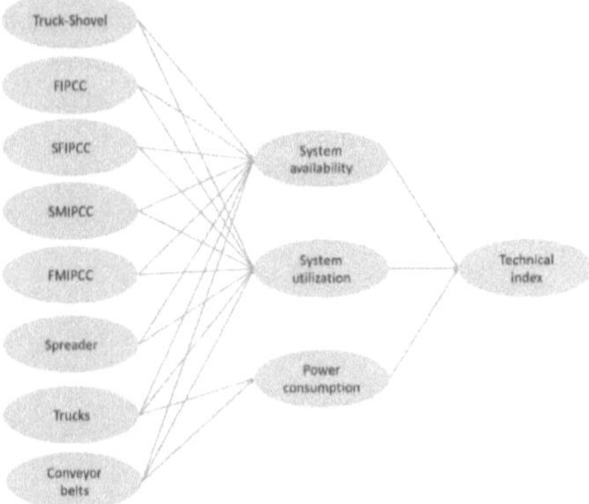

Figure 3-3. Cause tree of technical index

3.7.1. System availability

Availability for mining equipment considers any loss time that encompasses any activities in a stoppage in the planned production for a while, such as nonscheduled time, scheduled maintenance time, setup and adjustment time, etc. [61]. Accordingly, the availability equation can be described as follows [61]:

$$\text{Availability} = \frac{\text{Net available time} - \text{Downtime losses}}{\text{Net available time}} \qquad \text{Equation 3-1}$$

In terms of the available hour per year, the following equation can be defined:

$$\text{Available hour per year} = 8760 - \text{unplanned maintenance stoppages} \qquad \text{Equation 3-2}$$

However, there are different approaches that the availability of a system can be evaluated. The most important of them are as follows:

3.7.1.1. Serial configuration

In this type, it is considered that all the parts of the system are connected in a serial format, which all the components must be available for the system's availability [62]. The total availability results by multiplying of individual components' availabilities (Equation 3-3) [62]. Figure 3-5 shows a serial configuration of components in a system. The overall availability for this system is represented in Equation 3-4.

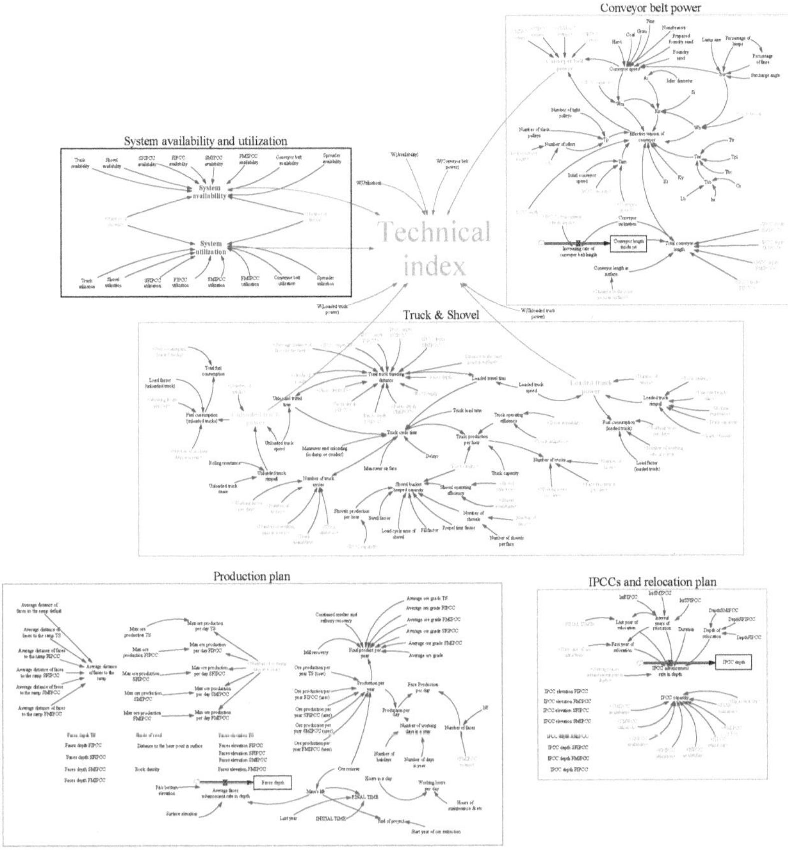

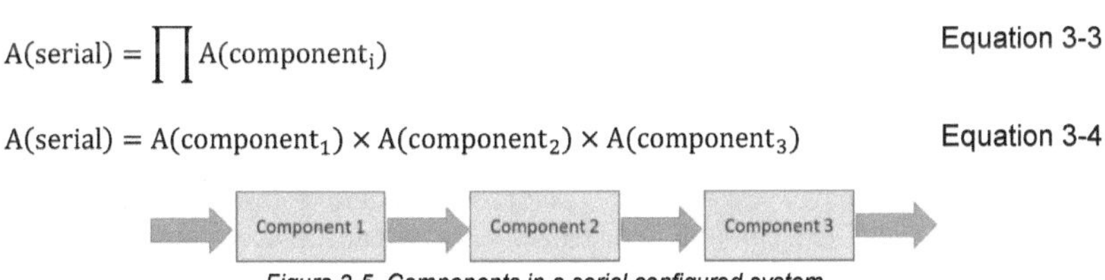

Figure 3-4. The general sketch of the system dynamics model built for the technical index in Vensim

$$A(\text{serial}) = \prod A(\text{component}_i)$$

Equation 3-3

$$A(\text{serial}) = A(\text{component}_1) \times A(\text{component}_2) \times A(\text{component}_3)$$

Equation 3-4

Figure 3-5. Components in a serial configured system

3.7.1.2. Parallel configuration

In this type, the whole system's availability is as far as one of the system's components is available [62]. Figure 3-6 shows a parallel configured system with three components.

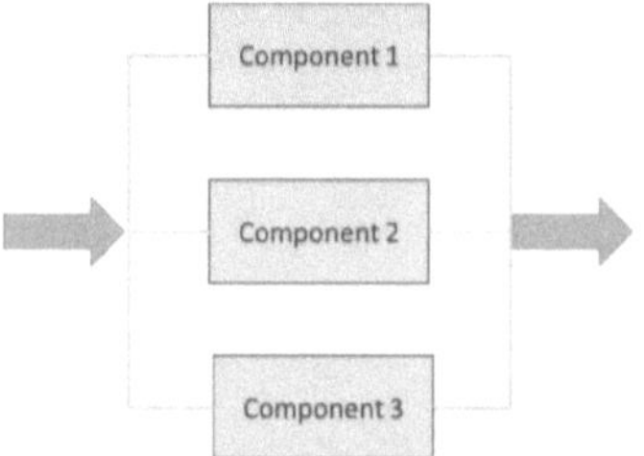

Figure 3-6. Components in a parallel configured system

In contrast with the serial configuration system, which the system works when all the components are working, this system fails if all the components fail [62]. The system's overall availability in this type of configuration is calculated through Equation 3-5. Equation 3-6 is an example of the total availability for a three-component system.

$$A(\text{parallel}) = 1 - \text{unavailability (parallel)}$$

Equation 3-5

$$= 1 - \prod [(1 - A(\text{component}_i))]$$

$$A(\text{parallel}) = 1$$

Equation 3-6

$$- ([1 - A(\text{component}_1)] \times [1 - A(\text{component}_2)]$$
$$\times [1 - A(\text{component}_3)])$$

3.7.1.3. Hybrid configuration

When a system is consists of many components that some of them have series, and the rest have a parallel configuration, this system has a hybrid configuration [62]. In this case, the parallel parts should be calculated together, and the result is counted and added as a block to the system. This process is continued until the availability of the whole system is determined. Figure 3-7 shows an example of a hybrid configuration in a system.

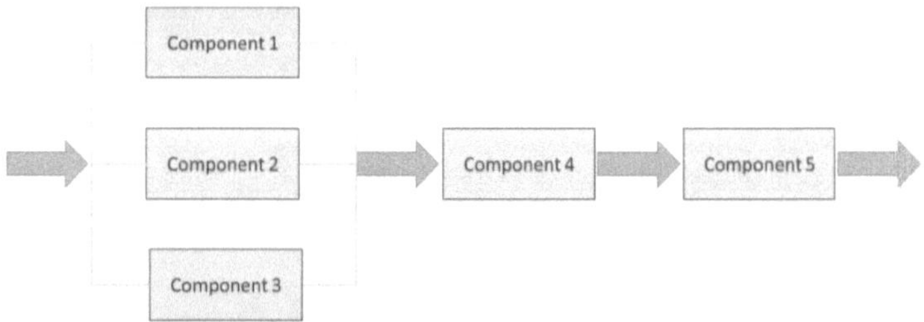

Figure 3-7. Hybrid configuration of a system

The Truck-Shovel system only includes trucks and shovels and considered as a discontinuous system [63]. Four different states can be considered to determine the Truck-Shovel system's availability, which is represented in Figure 3-8 and Table 3-1. Although State 1 and State 3 can be unrealistic, it was mentioned to represent all possible states. It is assumed that the crusher, conveyor belt, and spreader/stacker, which are located outside the pit in this system, have no effect on the availability of the Truck-Shovel system; because if each part of the crushing unit (i.e., crusher, conveyor belt, and spreader/stacker) is not available, the material can be depot in the stockpile for future crushing and processing.

Table 3-1. Four states of Truck-Shovel system

State	NS	NT	System configuration	System availability
1	1	1	Series	$A_S \times A_T$
2	1	>1	Hybrid	$A_S \times [1 - (1 - A_T)^{NT}]$
3	>1	1	Hybrid	$[1 - (1 - A_S)^{NS}] \times A_T$
4	>1	>1	Hybrid	$[1 - (1 - A_S)^{NS}] \times [1 - (1 - A_T)^{NT}]$

NS: Number of shovels
NT: Number of trucks
A_S: Shovel availability
A_T: Truck availability

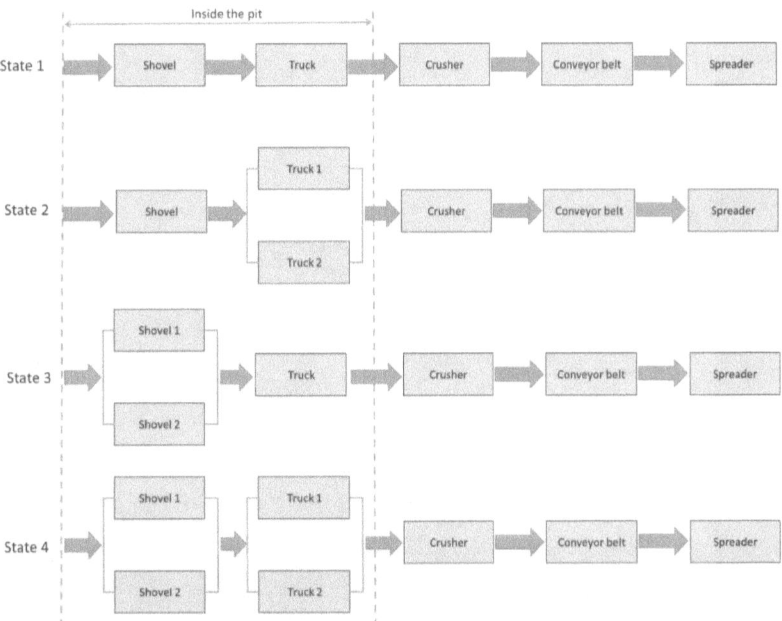

Figure 3-8. Schematic view of four states of the Truck-Shovel system

In the FIPCC, SFIPCC, and SMIPCC systems, the states are the same as the Truck-Shovel system, with one difference that the crusher and some parts of the conveyor belt are located inside the pit (Figure 3-9). Additionally, the crusher, conveyor belt, and spreader/stacker are acting as a continuous system, which the failure in each of them will fail the whole system of the FIPCC, SFIPCC, or SMIPCC [7]. The system availability for each state can be calculated based on the equations in Table 3-2. State 4 is assumed in modeling.

In the FMIPCC system, in which there are no trucks and the mobile crusher is directly fed by the shovels, two different states can be described as represented in Figure 3-10 and Table 3-3. This project focuses on State 1.

Table 3-2. Four states of FIPCC, SFIPCC and SMIPCC systems

State	NS	NT	System configuration	System availability
1	1	1	Series	$A_S \times A_T \times A_{IPCC} \times A_{CB} \times A_{Sp}$
2	1	>1	Hybrid	$A_S \times [1 - (1 - A_T)^{NT}] \times A_{IPCC} \times A_{CB} \times A_{Sp}$
3	>1	1	Hybrid	$[1 - (1 - A_S)^{NS}] \times A_T \times A_{IPCC} \times A_{CB} \times A_{Sp}$
4	>1	>1	Hybrid	$[1 - (1 - A_T)^{NT}] \times [1 - (1 - A_S)^{NS}] \times A_{IPCC} \times A_{CB} \times A_{Sp}$

A_{IPCC}: IPCC availability
A_{CB}: Conveyor belt availability
A_{Sp}: Spreader/stacker availability

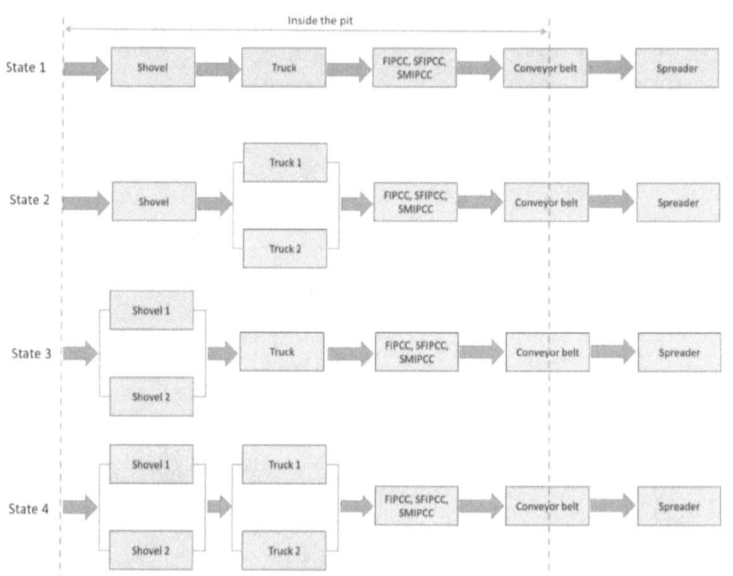

Figure 3-9. Schematic view of four states of the FIPCC, SFIPCC, and SMIPCC systems

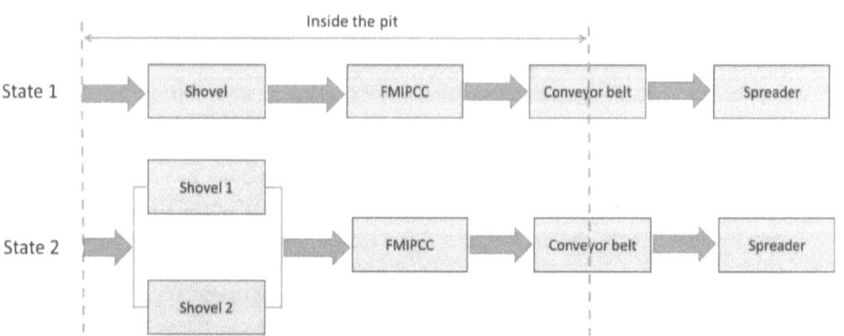

Figure 3-10. Schematic view of four states of the FMIPC system

Table 3-3. Four states of FMIPCC system

State	NS	System configuration	System availability
1	1	Series	$A_S \times A_{IPCC} \times A_{CB} \times A_{Sp}$
2	>1	Hybrid	$[1 - (1 - A_S)^{NS}] \times A_{IPCC} \times A_{CB} \times A_{Sp}$

3.7.2. System utilization

Utilization for any equipment is described as the total available time except the stoppage times, e.g., shift changes, meal breaks, etc. the following equation represents the utilized hour per year of a component as a function of availability:

$$\text{Utilized hour per year} = \text{Available hour} - (\text{planned} + \text{unplanned})\text{process delays}$$

Equation 3-7

Accordingly, the total utilization of the system is dependent on its components. For series and parallel systems, utilization can be defined as the following equations:

$$U_{series} = \frac{\min\{A_i \times U_i\}}{A_{system}}$$

Equation 3-8

$$U_{parallel} = \frac{\sum_{i=1}^{n}(A_i \times U_i)}{n \times A_{system}}$$

Equation 3-9

FMIPCC system is considered as a series system, and the other transportation systems are assumed as parallel. Figure 3-11 shows the system dynamics model of system availability and utilization designed in Vensim.

3.7.3. Power consumption

Power consumption in the transportation systems interprets the power consumed by the transportation components to deliver material from the origin to the destination. Accordingly, the main components of the transportation systems, which are trucks and conveyor belts, need to provide sufficient rimpull and effective tension to handle transferring material properly. These factors will result in consuming fuels in trucks and electricity in conveyor belts, which affect the technical and economic part of the project. For estimating the power consumption of each transportation system, the following factors were calculated:

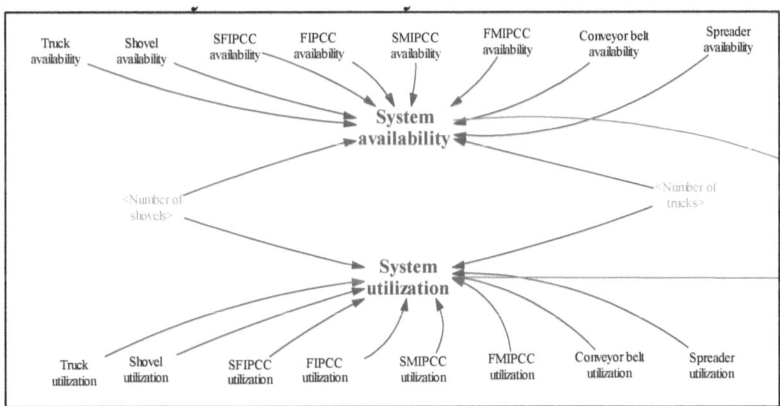

Figure 3-11. System dynamics model built in Vensim for system availability and utilization

3.7.3.1. Truck power

Equation 3-10 and Equation 3-11 represent the general relation for calculating the truck power in two modes of loaded and unloaded truck, respectively [64].

$$P_{\text{Loaded Truck}} = \frac{R_{LT} \times S_{LT}}{3.6} \cdot g \qquad \text{Equation 3-10}$$

$$P_{\text{Unloaded Truck}} = \frac{R_{UT} \times S_{UT}}{3.6} \cdot g \qquad \text{Equation 3-11}$$

R_{LT} and R_{UT} are the rimpull of the loaded and unloaded truck, S_{LT} and S_{UT} are the loaded and unloaded truck speed, respectively and, g is the gravitational acceleration. Rimpull equation for loaded and unloaded trucks can be calculated as the following equations [65]:

$$R_{LT} = (m_T + m_L)(R_r \pm G) \qquad \text{Equation 3-12}$$

$$R_{UT} = m_T(R_r \pm G) \qquad \text{Equation 3-13}$$

m_T, m_L, G, and R_r are the truck mass, load mass, road grade, and rolling resistance, respectively. For both loaded and unloaded trucks, the road grade could be plus or minus if drive uphill or downhill, respectively [66]. Figure 3-12 shows the system dynamics model built in Vensim for the loaded and unloaded truck power.

3.7.3.2. Conveyor belt power

As one of the major parts of the IPCC systems, the conveyor belt is responsible for the power consumption in the transportation systems. Accordingly, it is crucial to evaluate its contribution to this issue. One of the widely using equations for estimating the power of the conveyor belts is as follows [67]:

$$hp = \frac{T_e \times V}{33000} \qquad \text{Equation 3-14}$$

which T_e and V are belt tension in *lbs* and belt speed in *fpm,* respectively. Belt tension results from other tensions and forces, which constitutes Equation 3-15 [67]:

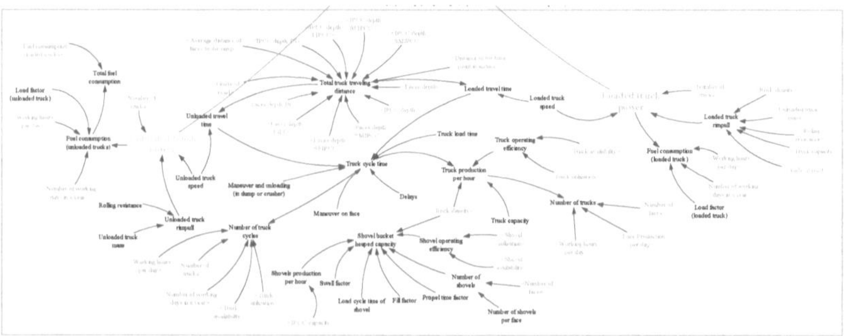

Figure 3-12. System dynamics model of loaded and unloaded truck power in Vensim

$$T_e = LK_t(K_x + K_y W_b + 0.015 W_b) + W_m(LK_y \pm H) + T_p + T_{am} + T_{ac} \qquad \text{Equation 3-15}$$

Which

- L is the length of the conveyor belt (ft),
- K_t is the ambient temperature correction factor,
- K_x is the factor used to calculate the frictional resistance of the idlers and the sliding resistance between the belt and idler rolls (lbs/ft),
- K_y is the carrying run factor used to calculate the combination of the resistance of the belt and the resistance of the load to flexure as the belt and load move over the idlers,
- W_b is the weight of the belt (lbs/ft),
- W_m is the weight of the material (lbs/ft),
- H is the vertical distance that material is lifted or lowered (ft),
- T_p is tension resulting from the resistance of the belt to flexure around the pulleys and the resistance of pulleys to rotation on their bearings, the total for all pulleys (lbs),
- T_{am} is the tension resulting from the force to accelerate the material continuously as it is fed onto the belts (lbs) and,
- T_{ac} is the total of the tensions stemming from the conveyor accessories (lbs).

While these parameters are directly related to the technical index, they are caused by different parameters and equations. All of them are represented in Appendix I. Figure 3-13 shows the built system dynamics model for the conveyor belt power in Vensim.

3.7.4. Technical index equation

To compare the different transportation systems in the technical index (TI), an equation was proposed based on the aforementioned parameters. This equation's logic is that each system with higher availability and utilization is preferable to the others. Besides, any transportation system with lower power consumption is better than the others. Since the importance of these items can be different, weighting factors can be applied as follows:

$$\sum_{a \in A} w(a) = 1 \qquad A = \{SA, SU, CBP, LTP, UTP\} \qquad \text{Equation 3-16}$$

In which SA, SU, CBP, LTP, and UTP are system availability, system utilization, conveyor belt power, loaded truck power, and unloaded truck power, respectively.

Accordingly, the following equation can be proposed:

$$TI = \frac{w(SA)\text{System availability} + w(SU)\text{System utilization}}{w(CBP)\text{Conveyor belt power} + w(LTP)\text{Loaded truck power} + w(UTP)\text{Unloaded truck power}} \qquad \text{Equation 3-17}$$

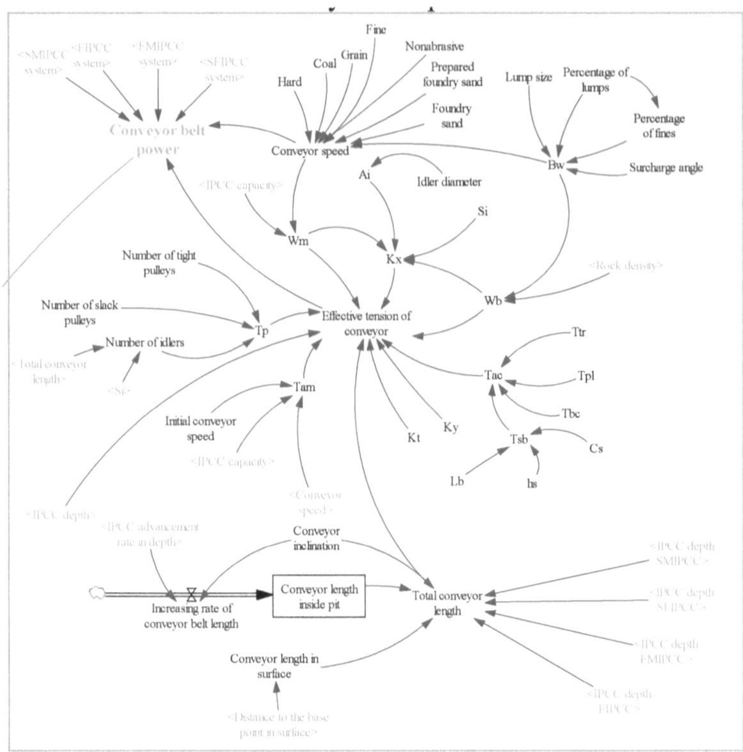

Figure 3-13. System dynamics model of the conveyor belt power

3.8. Economic index[5]

To define the economic index for the different transportation systems, the essential factors of income, total capital, and operating costs for each transportation system were determined, which will be explained in detail in the following sub-sections. Figure 3-14 shows the general sketch of the economic index in the system dynamics model.

3.8.1. Income

The general equation for the income rate is as Equation 3-18. Production per year of each transportation system could vary from one to another regarding its production plan. The final product price's future trend is estimated by the linear or exponential trend (Figure 3-15).

$$\text{Income rate} = \text{Final product price} \times \text{Final product per year} \qquad \text{Equation 3-18}$$

---

[5] This subsection (3.8) was published as a conference paper cited as:

Abbaspour, H., Drebenstedt, C. (2020). Determination of transition time from Truck-Shovel to an IPCC system considering economic viewpoint by system dynamics modeling. *In Topal, E. (Ed). Proceedings of the 28th International Symposium on Mine Planning and Equipment Selection – MPES 2019*, pp. 289-295.

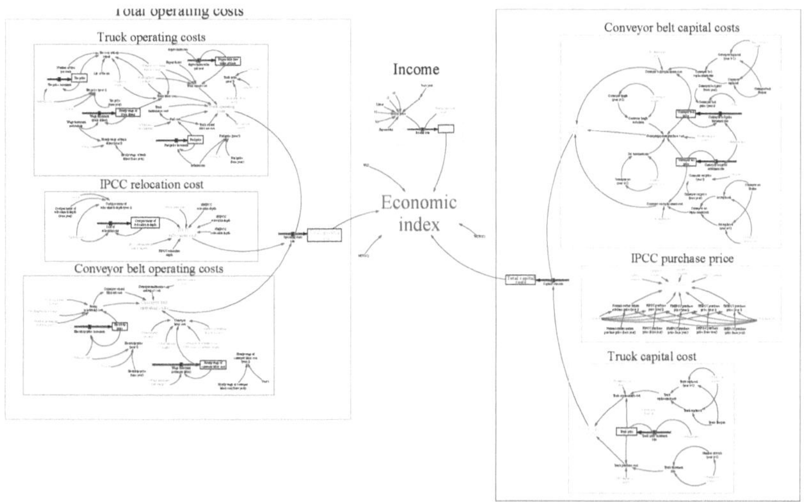

Figure 3-14. The general sketch of the economic index built in the system dynamics model in Vensim

3.8.2. Total capital costs

Total capital cost is constituted by summing the capital costs rate (conveyor belt and truck) and the investment cost for purchasing the IPCC system. Buying the IPCC system can occur once at the start of the project or any year after the project's start. However, the capital costs rate, which determines annual capital costs resulting from the sum of trucks and conveyor belts' capital costs, can occasionally occur through the mine life. The equations for the conveyor belt and truck capital costs are as follows:

Conveyor capital cost = Conveyor belt replacement cost +   Equation 3-19
Conveyor set replacement cost + Conveying system purchase cost

Truck capital cost = Truck purchase cost − Truck replacement cost   Equation 3-20

In the item "Truck purchase cost", the other items related to the trucks' ownership are hidden, e.g., insurance cost, import duties, interest, and investment opportunity cost. For trucks, a remaining value on the year of replacement by evaluating the depreciation rate is considered. Figure 3-16 depicts the system dynamics model for the total capital costs in the model.

3.8.3. Total operating costs

Total operating costs stem from three items: the conveyor belt operating cost, trucks operating costs, and the IPCC system's relocation cost. Each of the conveyor belt and trucks operating costs is calculated based on the following equations:

Truck operating cost = tire cost and repair + truck repair cost +   Equation 3-21
fuel cost + truck oil and lubricant cost + truck labor cost

Conveyor belt operating cost =   Equation 3-22
conveyor maintenance and repair cost + power(electricity)cost +
conveyor oil and lubricant cost + conveyor labor cost

All the other related equations for the income, operating, and capital costs of trucks and the conveyor belt are listed in Appendix II.

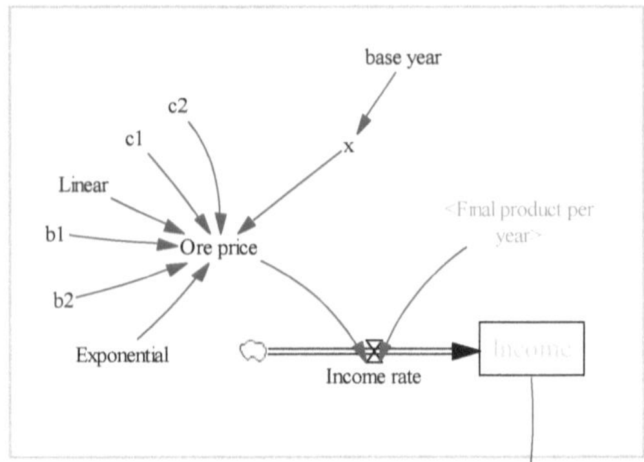

Figure 3-15. Sketch of the income in the system dynamics model

3.8.4. Economic index equation

Since each transportation system with the higher income and lower costs is preferable rather than others, economic index (EcI) is defined based on the portion of income to the total costs:

$$\text{EcI} = \frac{w(\text{I})\text{Income}}{w(\text{TCC})\text{Total capital costs} + w(\text{TOC})\text{Total operating costs}}$$

Equation 3-23

which $w(I)$, $w(TCC)$ and $w(TOC)$ are the weight factor of income, total capital costs, and total operating costs, respectively.

3.9. Environmental index construction[6]

The following items are taken into account for building the system dynamics model and measuring the environmental index of the system:

- Total emissions (CO_2, SO_2, and NO_x)
- Total particulate matter ($PM_{2.5}$, PM_{10}, and PM_{30})
- Total water consumption
- Equivalent noise level

---

[6] This subsection (3.9) was published as a conference paper cited as:

Abbaspour, H., Drebenstedt, C. (2019). Environmental comparison of different transportation systems – Truck-Shovel and IPCCs – in open pit mines by system dynamics modeling. *In Widzyk-Capehart, E., Hekmat, A., Singhal, R. (Eds). Proceedings of the 27th International Symposium on Mine Planning and Equipment Selection – MPES 2018, pp. 287-305.*

Each item is constituted from two individual items: the truck and conveyor belt, i.e., the total emissions, particulate matter, water consumption, and equivalent noise level depict the total emissions, particulate matter, water consumption, and equivalent noise level in trucks and conveyor belt. A general view of the system dynamics built for defining the transportation system's environmental index is shown in Figure 3-18. Each of the items mentioned above will be explained thoroughly in the following sub-sections.

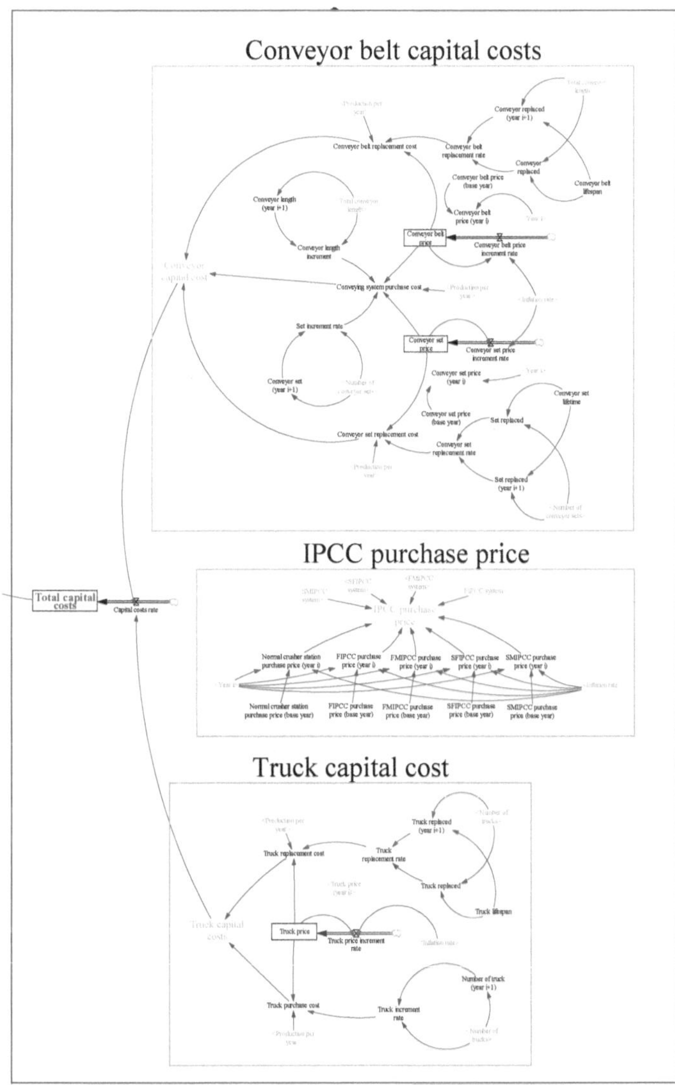

Figure 3-16. Total capital costs in system dynamics modeling

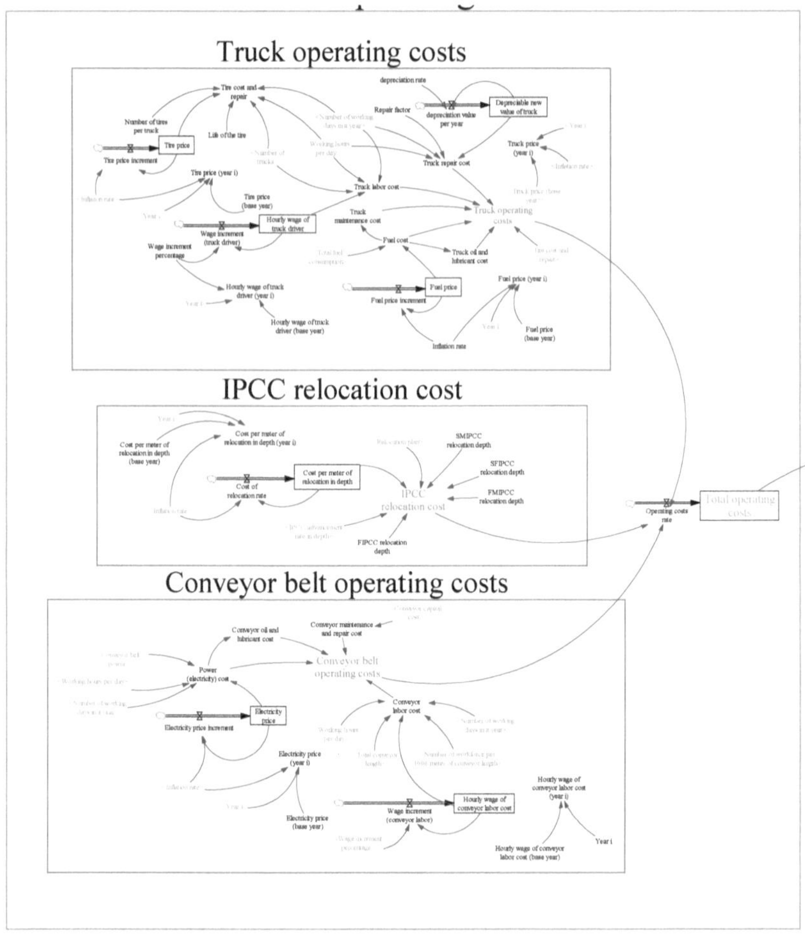

Figure 3-17. Total operating costs in system dynamics modeling

3.9.1. Total emissions (CO₂, SO₂, and NOₓ)

Trucks and conveyor belts, as the most important transportation equipment in the Truck-Shovel and IPCC systems, produce different emissions during the operation. Trucks and conveyor belts generate emissions by burning fossil fuels and electricity consumption generated from different energy sources (e.g., coal, oil, peat, and natural gas). This issue introduces the truck as a direct source of emissions and the conveyor belt as an indirect source of emissions. In this study, the most significant and known emissions are considered as CO_2, SO_2, and NO_x.

The emission factor is one of the most significant items that determines the quantity of the emission. However, it varies from one place to another [68, 69, 70]. Accordingly, one of the main tasks is to determine these factors. In this research, it is assumed that lignite is burnt in the power plant to produce electricity and diesel as the fuel of trucks. In this regard, the related emission factors for CO_2, SO_2, and NO_x are calculated (Table 3-4 and Table 3-5).

3.9.1.1.   Total emissions of conveyor belt

It is necessary to quantify the burnt fuel for producing electricity to calculate the conveyor belt emissions. To do so, the concepts of heat rate and heat content should be introduced. Heat rate is the energy consumed by a power plant or an electrical generator to produce one kilowatt-hour (kWh) of electricity [71], and heat content represents how much energy will be produced by burning a specific amount of fuel. Heat rate is expressed in Btu/kWh, and heat content is defined in Btu/t. By considering these explanations, the following equation for calculating the amount of burnt fuel is defined:

$$\text{Burnt fuel} = \frac{\text{Heat rate}}{\text{Heat content}} \quad \left(\frac{t}{kWh}\right) \qquad \text{Equation 3-24}$$

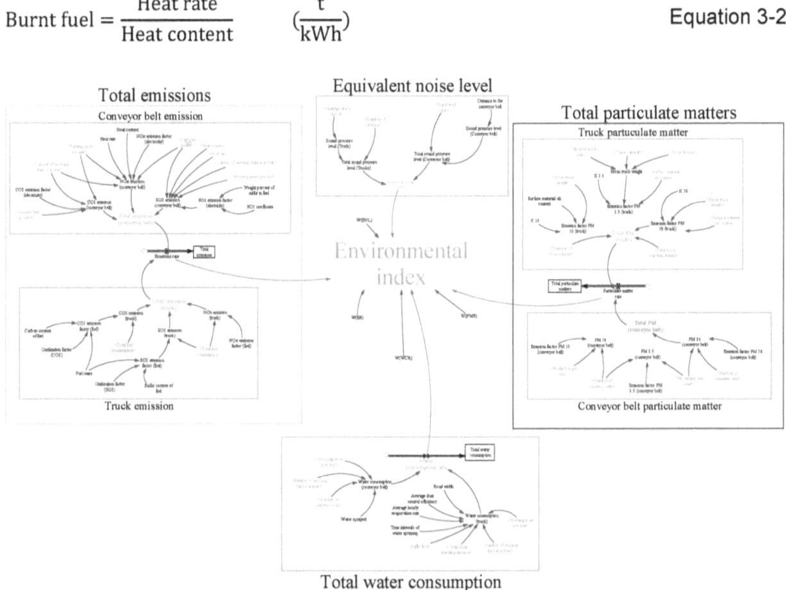

Figure 3-18. System dynamics model for the environmental index of the transportation system

Since it is assumed in this model that coal is burnt in power plants to produce electricity, the quantity of heat rate and heat content would be 10059 Btu/kWh and 21.258 MBtu/t, respectively [72, 73]. Regarding the emission factors of CO_2, SO_2, and NO_x (Table 3-4), the following equations for determining CO_2, SO_2, and NO_x can be defined:

CO_2 emission = CO_2 emission factor × Conveyor power × Number of working days in a year × Working hours per day

$$\text{Equation 3-25}$$

SO_2 emission = Burnt fuel × SO_2 emission factor × Conveyor power × Number of working days in a year × Working hours per day

$$\text{Equation 3-26}$$

NO_x emission = Burnt fuel × NO_x emission factor × Conveyor power × Number of working days in a year × Working hours per day

$$\text{Equation 3-27}$$

Table 3-4. Emission factors of CO_2, SO_2, and NO_x in conveyor belts

Emission factor	Equation/Quantity	Unit	Reference
CO_2	0.36	kg/kWh	[74]
SO_2	SO_2 coefficient × Weight percent of sulfur in fuel	kg/t	[75]
NO_x	6.81	kg/t	[75]

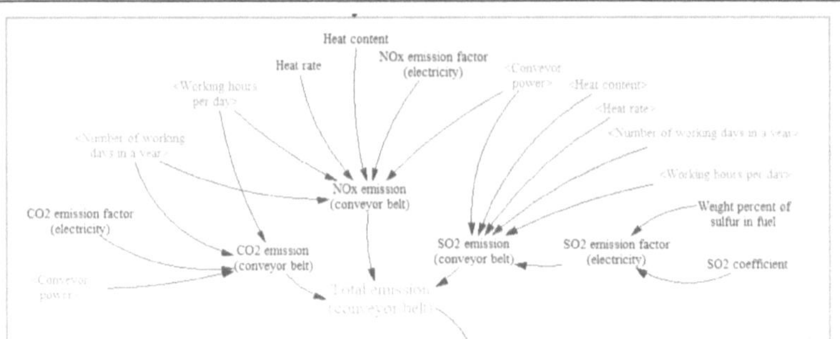

Figure 3-19. Total emissions from the conveyor belt in the system dynamics modeling

3.9.1.2.   Total emissions of trucks

The emission of trucks is originated from burning fuels in the trucks' engines. Accordingly, they are directly and by burning fuel are responsible for the emission in the mine site. On the contrary, the conveyor belt uses the electricity that is produced in the power plants. Accordingly, it is considered as an indirect agent of the emission. As the first step for evaluating emissions from trucks, it is required to determine the emission factors of CO_2, SO_2, and NO_x. For CO_2 and SO_2, the following equations are introduced based on the content of carbon and sulfur in diesel fuel [76]:

(44 / 12) × Oxidization factor (CO_2) × Carbon content of fuel × Fuel mass   Equation 3-28

(64 / 32) × Oxidization factor (SO_2) × Sulfur content of fuel × Fuel mass   Equation 3-29

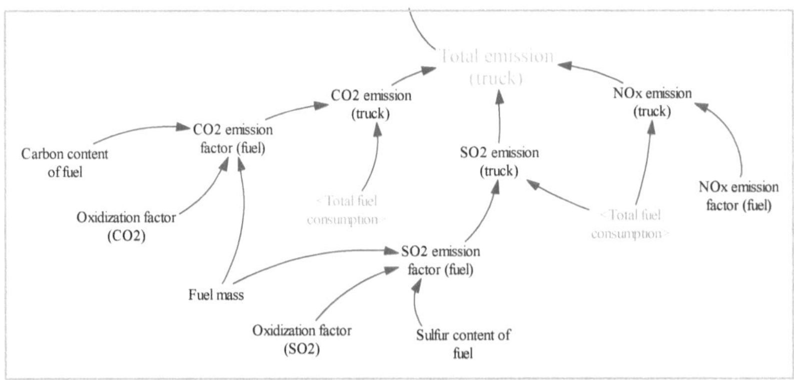

Figure 3-20. Total emissions from the truck in the system dynamics modeling

"44/12" and "64/32" representing the portion of the molecular weight of CO_2 and SO_2 to the molecular weight of carbon and sulfur, respectively. Oxidization factors depict how many carbon or sulfur percentages are transformed into CO_2 and SO_2 after burning, respectively. In fact, in the complete combustion, 100% of carbon and sulfur are burnt. However, it is not the case most of the time, and a percentage of carbon and sulfur remain unburnt. Therefore, the oxidization factors for CO_2 and SO_2 are considered as 99% and 98% in this study. The carbon and sulfur content of the fuels differ from one to another. However, in diesel fuel, they are generally counted as 86% and 15 ppm of the fuel mass, respectively. Fuel mass (diesel mass) is also set as 840 gr/L.

By multiplying the emission factors of CO_2, SO_2, and NO_x (Table 3-5) in trucks' total fuel consumption, the relevant emissions will be determined. The truck's total emissions will be the sum of the emissions of CO_2, SO_2, and NOx.

Table 3-5. Emission factors of CO_2, SO_2, and NO_x in trucks

Emission factor	Quantity	Unit	Reference
CO_2	2.622	kg/L	[76]
SO_2	2.47×10^{-5}	kg/L	[76]
NO_x	0.034	kg/L	[77]

3.9.2. Total particulate matters (PM$_{2.5}$, PM$_{10}$, and PM$_{30}$)

Particulate matter (PM) defines a mixture of solid and liquid particles scattered into the surrounding air [78]. Although different classification forms for particulate matter, the most recognized classification forms are categorizing by their physical size. Particle size is generally based on the aerodynamic diameter [78]. The abbreviation PM_x denotes all particles with a diameter of less than x micrometers. The most common PM_x, which are serious concerns in research, are $PM_{2.5}$, PM_{10}, and PM_{30}.

In the transportation system of a mine, various sources of particulate matter can be recognized. For instance, the transition points between conveyor sets (chutes) and wind erosion in transferring material by conveyor belt and the particulate matters from haul roads and wind erosion in moving material by trucks.

3.9.2.1.   Particulate matter generated from the conveyor belt

In the crushing process, various factors can affect particulate matter emission, such as rock type (ore or waste type), feed size and distribution, moisture content, output rate, crusher type, size reduction ratio, and fines content [79]. Different emission factors of the particulate matter are provided in any crushing steps through a variety of references [79]. Nevertheless, the conveyor belt's transition points (chutes) as the most crucial source in generating the particulate matters are considered. These emission factors, which can be measured and modified for any individual project, are shown in Table 3-6. The following equation can be set for calculating the total amount of the particulate matter 2.5, 10, and 30 based on the production rate:

Emission of ($PM_{2.5}$, PM_{10}, PM_{30}) = Number of conveyor sets ×   Equation 3-30

Production per year × Emission factor ($PM_{2.5}$, PM_{10}, PM_{30})

Table 3-6. Emission factors of particulate matter in the transition points of the conveyor belt

PM_x	Quantity (gr/t)	Reference
$PM_{2.5}$	6.5×10^{-6}	[79]
PM_{10}	2.3×10^{-5}	[79]
PM_{30}	7×10^{-5}	[79]

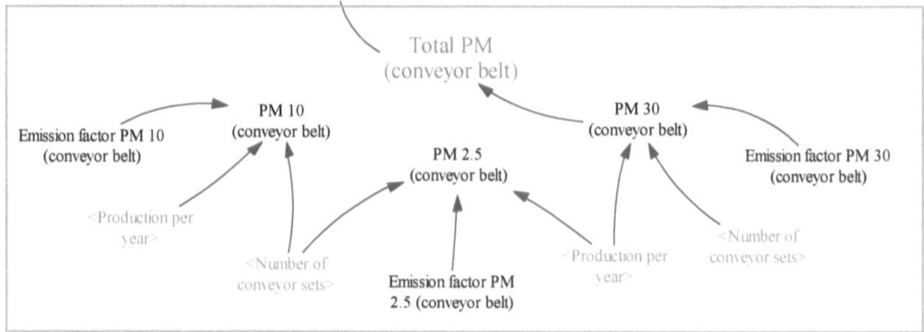

Figure 3-21. Total particulate matters from the conveyor belt in the system dynamics modeling

3.9.2.2. Particulate matter generated from trucks

In the mine sites, trucks are generally moving on roads and ramps that are unpaved. It causes dust generation that surface materials are pulverized when a truck travels on the road due to the trucks' wheel forces. Consequently, these powder materials lifted and dropped continuously by rolling wheels. In addition, a turbulent is generated behind the truck, which worsens the situation [80]. As previously described, the emission factor is an essential part of calculating the particulate matter. For trucks that are traveling on the unpaved surfaces at the industrial sites, the following emission factor is defined [80]:

$$\text{Emission factor of PMx} = k\left(\frac{s}{12}\right)^a \left(\frac{W}{3}\right)^b \qquad \text{Equation 3-31}$$

Which s is the surface material silt content (%), W is the mean vehicle weight (t), k is the emitted particulate matter per vehicle mile traveled[7] (lbs. /VMT) and a and b are constants (Table 3-7).

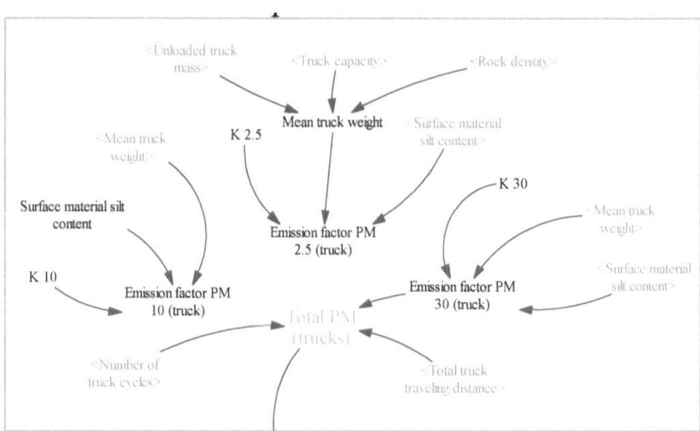

Figure 3-22. Total particulate matters from the trucks in the system dynamics modeling

---

Based on the definition of PM_x, PM_{30} encompasses both PM_{10} and $PM_{2.5}$. Accordingly, the total emission of the particulate matter of trucks can be estimated from the following equation:

Total PM = Emission factor PM_{30} × Number of cycles × Total truck traveling distance × 2

Equation 3-32

Table 3-7. The constant of emission factor of the particulate matters from trucks in the industrial roads [80]

Constant	$PM_{2.5}$	PM_{10}	PM_{30}
k (lbs. /VMT)	0.15	1.5	4.9
a	0.9	0.9	0.7
b	0.45	0.45	0.45

3.9.3. Total water consumption

In the transportation system of a mine, there are different means of water consumption. In trucks and conveyor belts, water is mostly consumed for the dust suppression caused by the trucks' movement and in transition points at the conveyor belts. However, there are other water consumption sources, e.g., the cooling system of trucks and washing. The water consumption in this model is considered the total water consumed for the particulate matter suppression caused by the trucks and conveyor belt.

Based on an empirical method [81], the control efficiency of spraying water is as the following equation:

$$C = 100 - \left(\frac{0.8 \times p \times d \times t}{i}\right)$$

Equation 3-33

Which p is the potential average hourly daytime evaporation rate (mm/h), d is the average hourly daytime traffic rate (vehicles/h), i is the application intensity of water (lit/m^2), and t represents the time between watering applications (h).

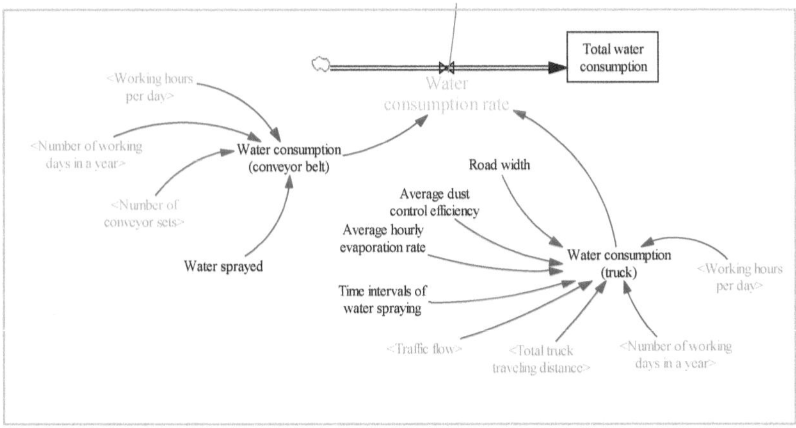

Figure 3-23. Total water consumption built in the system dynamics modeling

The water needed in the conveyor belt in order to suppress the dust is calculated from the following equation:

Water sprayed (lit/min) × Number of working days in a year × Working   Equation 3-34
hours per day × Number of conveyor sets × 60

3.9.4. Equivalent noise level

Making noise during the operation is one of the environmental issues that the transportation systems in the mines are responsible for. Not only will this affect the environmental condition of the mine site but also the health of the employees. Accordingly, this item can be evaluated in each type of transportation system. For this, the sound pressure level of each truck and conveyor belt are calculated and finally, an equivalent noise level for each transportation system was defined.

3.9.4.1.   Sound pressure level (Truck)

For measuring the noise level of any source, the concept of sound pressure level[8] is generally used. Accordingly, it is needed that this quantity to be calculated at a distance from the source.

For estimating trucks' noise, the data from a report related to different vehicles' noise emissions, including heavy-duty trucks on Germany's roads, is adopted [82]. Figure 3-24 shows the maximum sound pressure level measured for the heavy-duty vehicles with a power value of more than 250kW as a function of the vehicle's speed for the free-flowing traffic and the accelerating vehicles. The measuring conditions are based on the ISO 11819-1 (measuring distance 7.5 m from the centerline of the driving line and the height of 1.2 m above the road surface) [82].

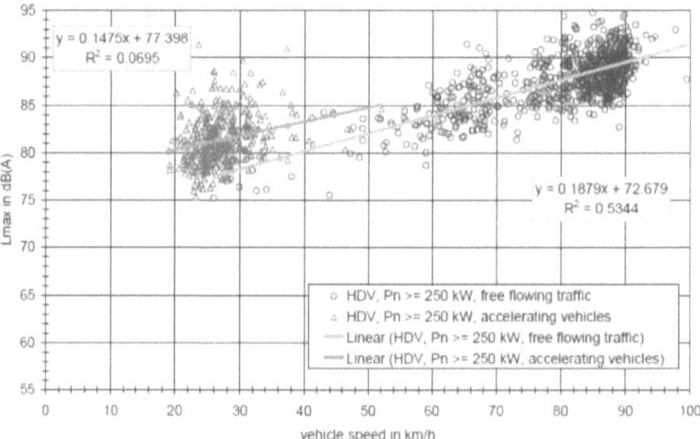

Figure 3-24. The maximum sound pressure level for the heavy-duty trucks with more than 250kW power in the free-flowing traffic and the accelerating vehicle [82]

Due to the lack of such data for mining projects, these data are used.

By considering the average truck speed and the free-flowing traffic, the following equation for estimating the sound pressure level of the trucks is taken into the model:

Sound pressure level (Truck) = 0.1879 × Average truck speed + 72.679   Equation 3-35

---

[8] Sound pressure level is produced in a certain distance of a source, which produce an amount of sound power level [131]

Since this equation is for a single truck, the total sound pressure level of all trucks can be determined as [83]:

$$\sum L_{\text{Trucks}} = \sum_{i=1}^{n} 10 \log\left(10^{\frac{L_i}{10}}\right) \qquad \text{Equation 3-36}$$

In which L_i ($i = 1, 2, \dots, n$) is the sound pressure level of each truck. Since this amount is assumed the same for all trucks, Equation 3-36 can be modified as follows:

$$\sum L_{\text{Trucks}} = \text{Number of trucks} \times 10 \log\left(10^{\frac{L_1}{10}}\right) \qquad \text{Equation 3-37}$$

3.9.4.2.   Sound pressure level (Conveyor belt)

The noise resulted from the conveyor belt can be a function of different factors, e.g., the belt speed, belt capacity, type of idlers, etc. However, the sound pressure level of the noises caused by the conveyor belt differs based on the distance to the measuring point. Table 3-8 can be concluded for the relation between sound pressure level and the distance to the conveyor belt, which is based on the relative works.

By taking the standards of measurement in the sound power pressure level of trucks, which the measurement point should be located at a distance of 7.5 m and the height of 1.2 m, the first four items of Table 3-8 are taken. Accordingly, the relation between sound pressure level and the distance to the source can be shown in Figure 3-25.

Since these tests are performed for a single idler, with the same paradigm that explained in the sound pressure level for trucks, the following equation is used for determining the total sound pressure level of all idlers:

$$\sum L_{\text{Idlers}} = \text{Number of idlers} \times 10 \log\left(10^{\frac{L_1}{10}}\right) \qquad \text{Equation 3-38}$$

The system dynamics model for the equivalent noise level is shown in Figure 3-26

Table 3-8. Sound pressure level measurement for conveyor belts [84, 85]

Sound pressure level (dB)	Height (m)	Distance (m)	Belt speed (m/s)	Capacity (t/h)	Idler type	Reference
79	1.2	1	6.55	1000	standard	[84]
69	1.2	10	6.55	1000	standard	[84]
59	1.2	100	6.55	1000	standard	[84]
59	1.2	1000	6.55	1000	standard	[84]
85*	1.5	3	5	10000	standard	[85]
84*	1.5	3	5	10000	standard	[85]
86*	1.5	3	5	10000	standard	[85]

*these quantities were obtained after three times repeating the test

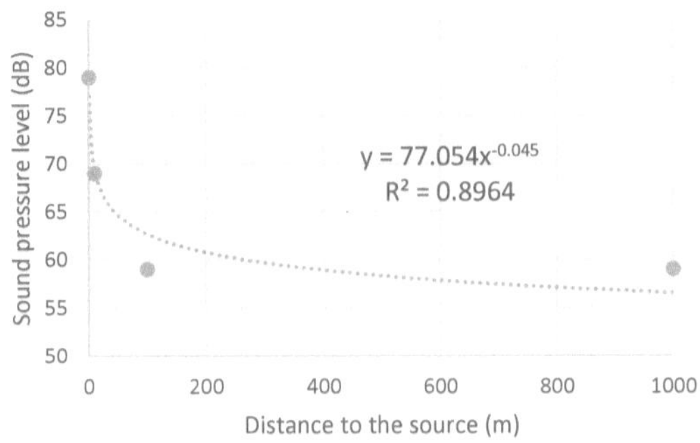

Figure 3-25. Sound pressure level as the function of the distance to the source

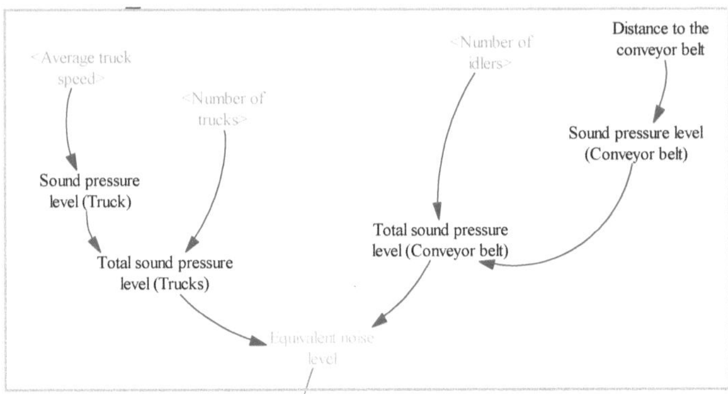

Figure 3-26. System dynamics model of the equivalent noise level

Finally, the equivalent sound pressure level will be estimated by the following equation:

$$\sum L_{Equivalent} = 10\log\left(10^{\frac{\sum L_{Trucks}}{10}} + 10^{\frac{\sum L_{Idlers}}{10}}\right)$$

Equation 3-39

3.9.5. Environmental index

As mentioned before, the environmental factors that define the transportation system's environmental index are emission rate, particulate matter rate, water consumption rate, and equivalent noise level. The relation of the environmental index with each of these factors is reversed, in which the higher emissions, particulate matter, water consumption, and noise level result in a more unfavorable environmental situation. Accordingly, the environmental index (EI) can be defined as the following equation:

$$EI = \frac{c}{w(ER)\text{Emissions rate} + w(PMR)\text{PMs rate} + w(WCR)\text{Water consumption rate} + w(ENL)\text{Equivalent noise level}}$$

Equation 3-40

In which c is a constant and $w(ER), w(PMR), w(WCR)$ and $w(ENL)$ are the weighting factors of emissions rate, PMs rate, water consumption rate, and equivalent noise level, respectively.

3.10. Safety and social indexes construction[9]

As mentioned in Chapter 2, the first step in building a causal loop model is defining constants and variables, which determine the system behavior and interactively influence each other and the whole system. Since the Truck-Shovel and IPCC systems are constituted from trucks and conveyor belts, which the former affects the safety of roads and the latter impact the employees' safety, the most relevant and measurable factors in determining the safety level of these transportation systems are considered. In this regard, the following items are taken into account to build the system dynamics model and measuring the safety and social index:

- Average truck speed
- Traffic flow
- Traffic density
- Training
- Lost time injury frequency rate
- Number of conveyor set
- Health index

In the safety part, the average truck speed, traffic flow, traffic density, and lost time injury frequency rate represent the roads' safety level, and the number of conveyor sets is assumed as the employees' safety. Whereas these variables have an inverse relation with the safety [86, 87, 88], training has a direct relation (Equation 3-49), in which the higher training results in the higher safety [89].

Additionally, the number of employees, training, and health index, which are more common as the social factors in the mining projects [90, 91] and can be quantified in the model, are defined for evaluating social index.

"Training" factor is defined in both safety and social indexes. This is an example of the interactive behavior of system dynamics modeling, which can simultaneously affect the different parts of the system. Figure 3-27a shows the first causal loop diagram, which is developed for defining the safety and social indexes.

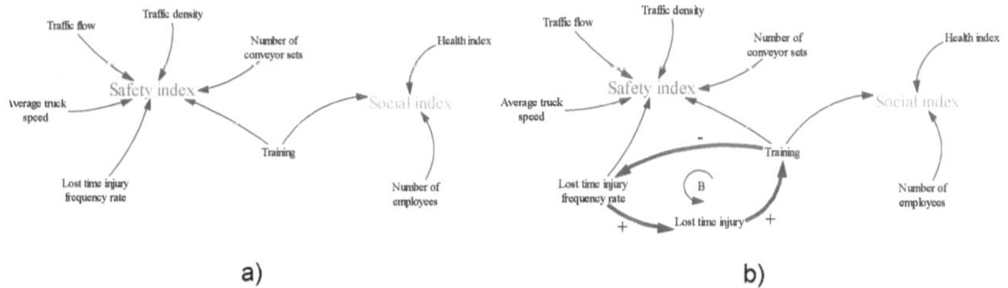

---

[9] This subsection (3.10) was published as a journal paper cited as:

Abbaspour, H., Drebenstedt, C., Dindarloo, S. R. (2018). Evaluation of safety and social indexes in the selection of transportation system alternatives (Truck-Shovel and IPCCs) in open pit mines. Safety Science 108, 1-12.

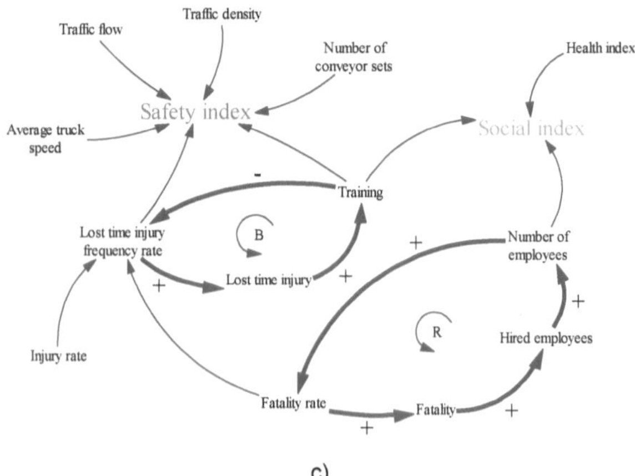

Figure 3-27. The causal loop diagram of defining safety and social indexes a) first causal loop diagram b) causal loop diagram with a balancing feedback c) the final causal loop diagram with one balancing and one reinforcing feedback

It is considered that the lost time injury frequency rate decreases if the training increases. In addition, when the lost time injury frequency rate increases, more training is needed to improve the level of employees' knowledge about safety. This description leads to balancing feedback (B) (Figure 3-27b). If the number of employees increases, the fatality rate caused by the incidents during the work increases likewise. Accordingly, for compensating the labor force, it would be necessary that the new employees be hired. It constitutes a reinforcing feedback loop (R) (Figure 3-27c).

The final stock-flow model for the safety and social indexes can be constructed (Figure 3-28). In the following subsections, each variable and its relevant quantity or equation are defined.

3.10.1.　Average truck speed

Based on the literature, the vehicles' speed is considered as an effective parameter in the roads' safety through happening accidents, fatalities, and injuries [92, 88, 93, 94, 95, 96] and, it is considered as one of the determining factors of safety index. Since the truck's speed is different while loaded and unloaded, the average truck speed is assumed as the average of unloaded and loaded truck speed. This factor will be zero for the FMIPCC system due to not operating trucks in this system.

3.10.2.　Traffic flow and traffic density

Traffic is one of the critical items that directly affect the safety of the mine's road and impacts the transportation system. There are many studies on traffic issues on the roads and highways [87, 97, 98, 99, 100, 101, 102]. However, the number of research for describing the mining roads' safety level is few [103, 104]. The typical traffic items that affect the road's safety are traffic density and traffic flow. The definitions of these factors are:

> ➤ Density: "The number of vehicles occupying a section of the roadway in a single line" [105]. The following equation is set for this variable in the model:

$$\text{Traffic density} = \frac{\text{Number of trucks}}{\text{Total truck traveling distance}}$$

Equation 3-41

> Flow: "The number of vehicles passing a specific point or short section in a given period in a single line" [105]. The following equation is set for this variable in the model:

$$\text{Traffic flow} = \frac{\text{Number of trucks}}{\text{working hours}}$$

Equation 3-42

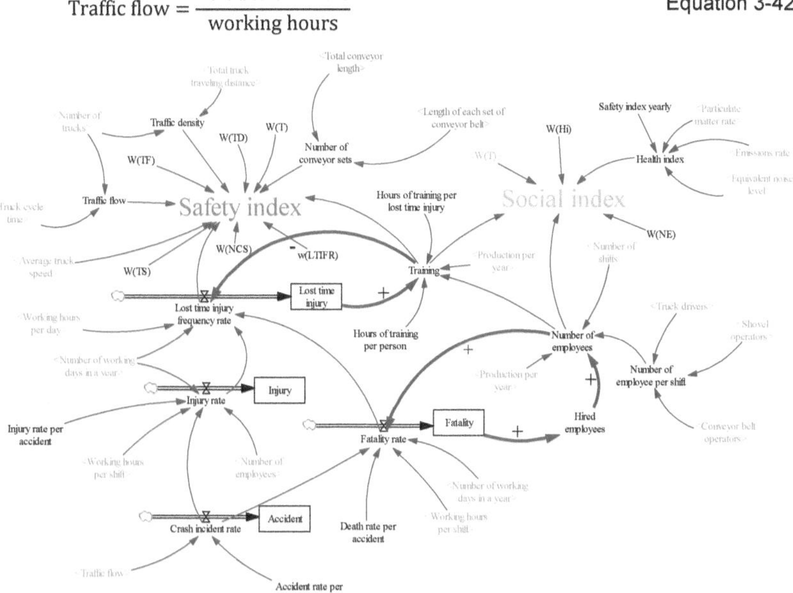

Figure 3-28. The system dynamics model for safety and social indexes

Since the loading and unloading of trucks is a cycle (i.e., loading at the face, traveling to the destination, unloading, return to the face), the traffic flow represents how many travels are recorded in one hour by trucks in one direction. Hence, the unit of traffic flow would be travels/hour.

Speed, traffic density, and traffic flow have a negative impact on the safety index, such that the higher speed, higher traffic density, and traffic flow eventuate the lower safety index. Also, the number of accidents is defined as a function of the model's traffic flow, which individually affects the injury and fatality rate (Figure 3-28).

3.10.3. Number of conveyor sets

Based on the accident reports provided by Mine Safety and Health Administration (MSHA) between 1995 to 2015 [106], 31 cases of a total of 65 accidents (48%) happened in the moving parts of the conveyor belts (Table 3-9).

The statistics gathered based on the work done by the Health and Safety Executive department [107] of the United Kingdom between 1986/87 to 1990/91 in mine areas show 50% of fatalities, 52% of major injury and 56% of more than three days injury (rows 2 and 3 in Table 3-10).

Table 3-9. Number of accidents based on the location of occurrence in the conveyor belt [106]

Location of accident in the conveyor belt	Number of accidents
Head pulley	4
Drive unit	4
Tail pulley	9
Idler and pinch points	12
Tension roller	1
Tripper roller	1
TOTAL accidents related to location	31

Table 3-10. Conveyor belt accident by category in the UK mines between 1986/87 to 1990/91 [107]

#	Category	Fatal	%	Major injury	%	Over three days of injury	%
1	Inadequate clearances or guards	0	0	11	14	27	4
2	Maintenance, on or around moving or stalled conveyors	0	0	25	32	118	18
3	Maintenance, on or around stationary conveyors	1	50	15	20	262	38
4	Misuse of equipment	0	0	12	16	36	5
5	Blocked chutes, falling spillage	0	0	8	10	97	14
6	Use of conveyor as working platform	0	0	6	8	55	8
7	Use of conveyor to transport materials	1	50	0	0	32	5
8	Struck or fell while crossing conveyor	0	0	0	0	57	8
	TOTAL	2	100	77	100	684	100

These statistics show the remarkable role of the conveyor belts' moving parts resulting in different types of accidents. Since each set of a conveyor belt consists of one head pulley, one tail pulley, a driving unit, and idlers, it is proper to keep the number of conveyor sets as low as possible. These will be resulted in:

- ➢ reduction in the maintenance and cleaning work for the head pulley, tail pulley, driving units, feed, and discharge chutes.
- ➢ reduction in the safety guards, which are needed to be installed at these rolling parts. Accordingly, the costs will be decreased as well.
- ➢ easier controllability and inspection of these parts.
- ➢ the lower quantity of conveyer belt sets, which decreases the possibility of failure in the continuous system.

Accordingly, the "Number of conveyor sets" variable is placed in the model to measure the level of safety in the conveyor belts. The following equation for this variable is used in the model:

$$\text{Number of conveyor sets} = \frac{\text{Total conveyor length}}{\text{Length of each set of conveyor belt}} \qquad \text{Equation 3-43}$$

3.10.4.   Lost time injury frequency rate

Lost-time injury frequency rate is one of the factors that directly affect the safety level of an industry or mine [86, 108, 109]. It is defined as the total amount of fatalities, injuries, and permanent disabilities that happen while working [110]. Accordingly, the lost time injury frequency rate refers to the lost time injury occurred in a period [110] and can be calculated through the following equation:

$$\text{Lost time injury frequency rate} = \frac{\text{Injury rate} + \text{Fatality rate}}{\text{Total working hours in a year}} \times 10^6 \qquad \text{Equation 3-44}$$

This equation identifies the number of lost-time injuries per one million hours of working. For calculating injury and fatality rates, a statistical analysis was carried out based on the injury and fatality rate data of Mine Safety and Health Administration (MSHA) in the US mines between 1995 to 2015 [106]. Assuming 1995 as the base year (year zero), the trend of the fatality and injury rates per 200,000 hours of working will be as Figure 3-29.

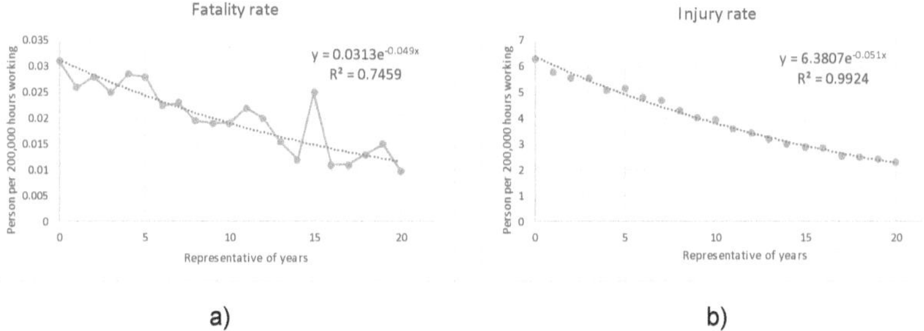

Figure 3-29. Data and trend of a) fatality rate and b) injury rate between 1995 to 2015 in the US mines

Injury and fatality rates are set as the average number of injuries and fatalities resulted during working hours based on the evaluated trends (Figure 3-29) and the number of accidents caused by trucks as a function of traffic flow (Figure 3-28).

Notably, there is a positive feedback loop among "Fatality rate", "Fatality", "Hired employees", and "Number of employees". It means whenever the number of employees increases, the quantity of fatality also increases. In addition, for substituting the fatalities, it is needed for new employees to be hired, which lead to the increment of the number of employees (Figure 3-28).

3.10.5. Training

Regarding the safety rules in the mining sites, any employee needs to pass training periods [111]. This training depends on the employees' responsibility and activity. This model assumes that certain hours of training for each employee per year are required. Additionally, a certain amount of training hours per lost-time injury is assumed through reports, presentations, meetings, etc., to increase the employees' awareness. Therefore, the following equation for measuring the training in the mine sites is considered:

$$\begin{aligned}\text{Training} = &\text{ Number of employees} \times \text{Hours of training per person} \\ &+ \text{Hours of training per lost time injury} \times \text{Lost time injury}\end{aligned} \qquad \text{Equation 3-45}$$

There is a negative feedback loop among "Training", "Lost time injury frequency rate", and "Lost time injury", in which by increasing the lost time injury, the hours of training must be increased in order to reduce the probability of future accidents. Furthermore, by increasing the hours of training, the lost time injury frequency rate will be decreased in a portion of hours of training (Figure 3-28). Since the training contributes to the improvement of the level of education and awareness through learning new skills, information, and knowledge, this factor can also influence social index.

3.10.6.   Number of employees

Job creation is one of the social impacts of a mining project. By progressing the mining project, hiring new employees, especially in the transportation system, will increase. Therefore, by deepening the mine, more trucks or sets of conveyor belts are needed for transferring ore/waste to the crusher/waste dump. On the other hand, some fatalities might happen in the mine that will demand new employees to be hired. Hence, a positive feedback loop among "Fatality"; Hired employees", and "Number of employees" is designed in the model (Figure 3-28). The employees constitute the number of employees in the transportation section (truck drivers, shovel operators, and conveyor belt operators) in all shifts. Therefore, the following equations can be defined for determining the number of employees:

$$\text{Number of employees per shift} = \text{Truck drivres} + \text{Shovel operators} + \text{Conveyor belt operators}$$

Equation 3-46

$$\text{Number of employees} = \text{Number of employees per shift} \times \text{Number of shifts} + \text{Hired employees}$$

Equation 3-47

3.10.7.   Health index

Health index is defined as a function of safety index, particulate matter rate, emissions rate, and equivalent noise level so that the higher safety index, the better health index will be reached. On the contrary, more particulate matter rate, emissions rate, and noise level will result in a lower health level. Accordingly, the following equation is introduced for defining health index:

$$\text{Health index} = \frac{\text{Safety index}}{\text{Particulate matters rate} + \text{Emissions rate} + \text{Equivalent noise level}}$$

Equation 3-48

3.10.8.   Safety and social indexes

Safety and social indexes are defined according to their relation to the lost time injury frequency rate, number of conveyor sets, traffic density, traffic flow, average truck speed, number of employees, training, and health index. For instance, by increasing the training, the safety level will increase as well. On the contrary, by increasing the lost time injury frequency rate, the safety level will decrease. As a result, the safety index (SaI) and social index (SI) based on their direct or reverse relation to variables, are introduced. These indexes are used to compare different alternatives, i.e., the Truck-Shovel, FIPCC, SFIPCC, SMIPCC, and FMIPCC systems. These indexes can be described as follows:

Equation 3-49

$$\text{SaI} = \frac{w(T)Training}{w(LTIFR)Lost\ time\ injury\ frequency\ rate + w(NCS)Number\ of\ conveyor\ sets + w(TD)Traffic\ density + w(TF)Traffic\ flow + w(ATS)Average\ truck\ speed}$$

Equation 3-50

$$\text{SI} = w(T)Training + w(NE)Number\ of\ employees + w(HI)Health\ index$$

In which $w(T), w(LTIFR), w(NCS), w(TD), w(TF), w(ATS), w(NE)$ and $w(HI)$ are the weight factors of the training, lost time injury frequency rate, number of conveyor sets, traffic density, traffic flow, average truck speed, number of employees, and health index, respectively.

3.11. TEcESaS Indexes software[10]

As it was described in the previous sections, many inputs should be set into the model to run it and get the results. It will make it difficult for users to deal with the model efficiently. Additionally, the complex environment of the system dynamics model in Vensim may cause it confusing for users that want to work on it. Accordingly, designing and preparing software for the model is high in demand. This software will help users easily interact with the model like a technical software and does not need installation of the whole Vensim software and just installation of the "Vensim Model Reader"[11]. It will help users to open any Vensim model or Venapp freely [112]. It should be noted that although simulation through the Vensim Model Reader is possible, the sensitivity simulation cannot be handled by it.

This software is named "TEcESaS indexes" (Appendix III), which starts with a welcome page (Figure 3-30a), and by clicking the "Continue" button, the "Main Menu" (Figure 3-30b) will appear, and the software can be used. There are five other options in the "Main Menu" page, which will be explained in the following sub-sections.

3.11.1. Input Menu

By clicking on the "Inputs Menu" (Figure 3-31a), its relevant page including four other options, which are "Technical Inputs Menu", "Economic Inputs", Environmental Inputs", and "Safety and Social Inputs", will be opened. "Technical Inputs Menu" (Figure 3-31b) offers two sub-items of "General Technical Inputs" and "Production and Relocation Plan". In the "General Technical Inputs", all the constants that must be determined by the user are listed and can be simply entered into the related box (Figure 3-32a). These inputs are divided in nine categories of "Mine's Specification", "Truck's Specification", "Shovel's Specification", "IPCC's Specifications", Conveyor Belt's Specifications", "Spreader's/Stacker's Specification", "Material Specifications", "Mill Specification" and "weights". In the "Production and Relocation Plan", the user is asked to specify if there is any production or relocation plan for the project. If this is not the case, the software will be run in the default mode (Figure 3-32b). Otherwise, a specific Excel file called "ProductionRelocation.xlsx" should be filled that the software recalls the production and relocation plan data.

3.11.2. Simulation Menu

After importing the inputs, the model can be run. "Simulation Menu" includes model simulation in each mode of transportation, i.e., the Truck-Shovel, FIPCC, SFIPCC, SMIPCC, and FMIPCC or merely by clicking the "Simulate model in all modes" (Figure 3-34a), which will be run in all transportation systems. When the simulation is completed,

---

[10] This software was published in the Mendeley Dataset with the DOI 10.17632/b75sdckjg2.2

a "Run completed!" message will be appeared on the screen (Figure 3-34b). If the run is not completed successfully, an error message relevant to the cause of the error will appear.

The other pages of "Economic Inputs", "Environmental Inputs", and "Safety and Social Inputs" are depicted in Figure 3-33.

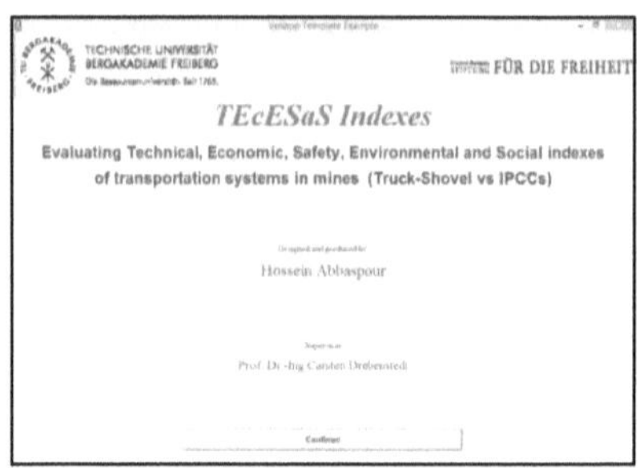

a)

b)

Figure 3-30. a) Welcome page and b) "Main Menu" of the TEcESaS Indexes software

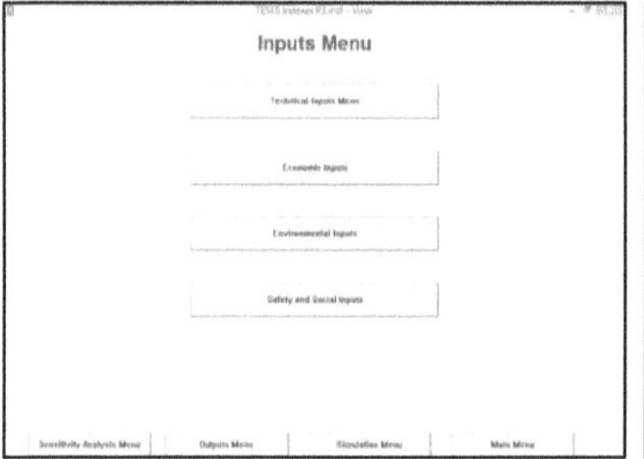
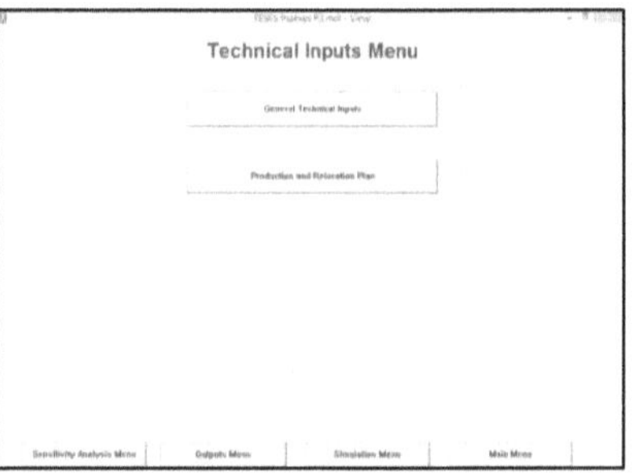

a)

b)

Figure 3-31. a) "Inputs Menu" and b) "Technical Inputs Menu" of the TEcESaS Indexes software

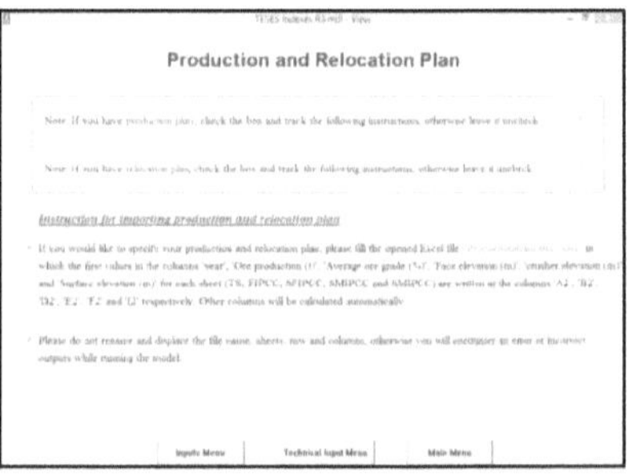

a)

b)

Figure 3-32. a) "General Technical Inputs" and b) "Production and Relocation Plan" in the TEcESaS Indexes software

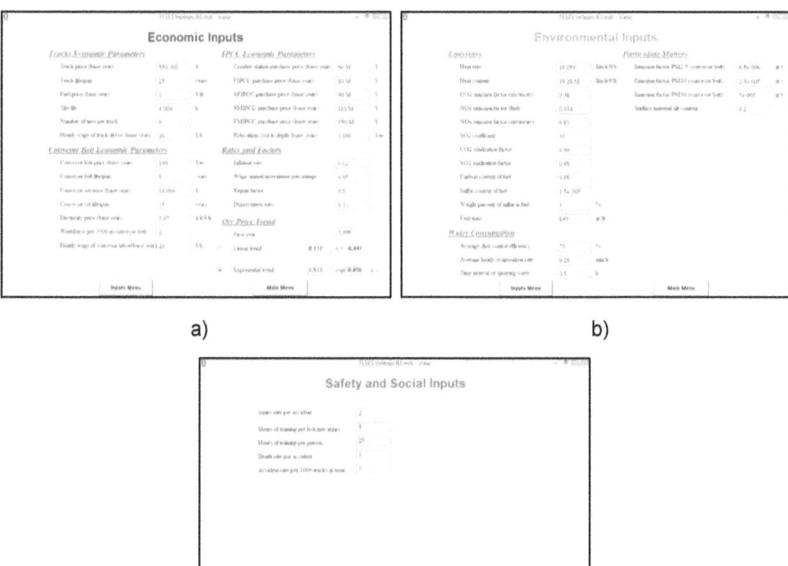

<table>
<tr><td align="center">a)</td><td align="center">b)</td></tr>
</table>

c)

Figure 3-33. a) "Economic Inputs", b) "Environmental Inputs" and c) "Safety and Social Inputs" in the TEcESaS Indexes software

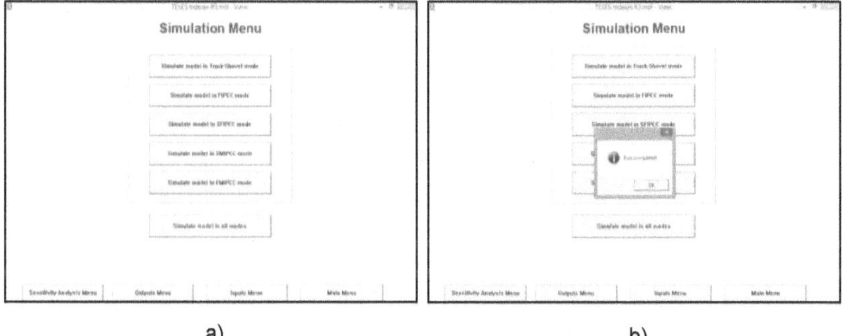

<table>
<tr><td align="center">a)</td><td align="center">b)</td></tr>
</table>

Figure 3-34. "Simulation Menu" a) before starting and b) after completing the simulation

3.11.3.　Outputs Menu

In the "Outputs Menu", users can select among the different options to observe the outputs, which could be in any individual technical, economic, environmental, safety and social outputs or any related index (Figure 3-35). For example, the first page of the "Technical outputs" and the "Technical index" is shown in Figure 3-36.

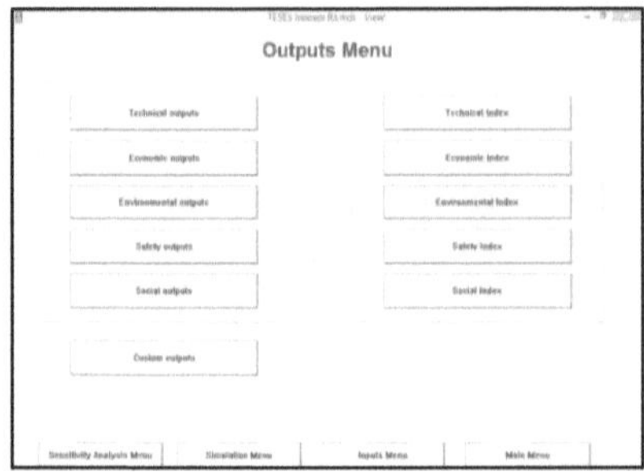

Figure 3-35. "Outputs Menu" of the TEcESaS Indexes software

There is also the "Custom output" in this menu that users can choose the variable they wish, and its related graph and table will be shown. For instance, the graph and table of the "Lost time injury" are shown in Figure 3-37.

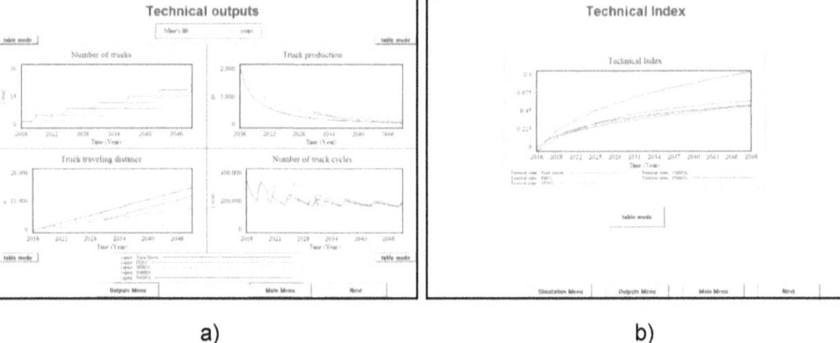

a)   b)

Figure 3-36. a) "Technical outputs" and b) Technical Index of the TEcESaS Indexes software

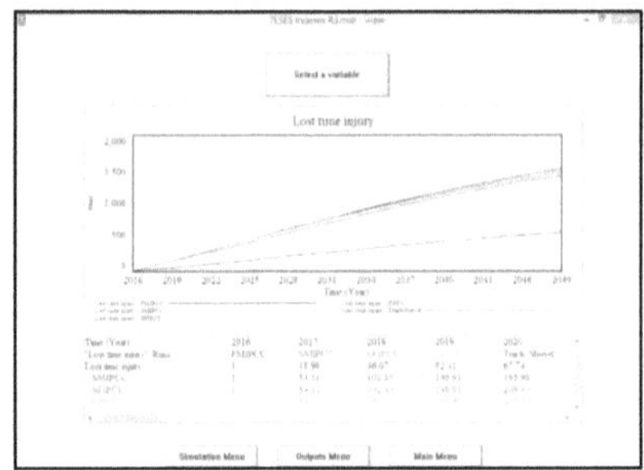

Figure 3-37. "Lost time injury" in the "Custom output" page

3.11.4.   Sensitivity Analysis Menu

In this menu, three parts, "Parameters", "Sensitivity runs", and "Sensitivity outputs" are designed (Figure 3-38). In the "Parameters" part, users can specify all of the technical, economic, environmental, safety, and social parameters and their related distribution based on their data. For instance, Figure 3-39 shows the technical parameters for running the sensitivity analysis, which is mentioned in Table 5-8.

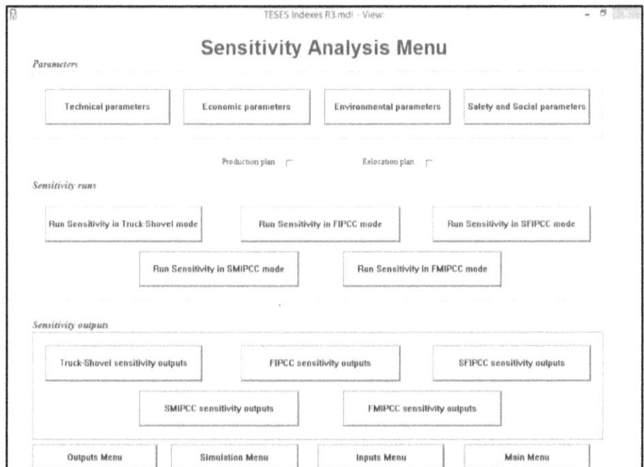

Figure 3-38. "Sensitivity Analysis Menu" in the TEcESaS Indexes software

By pressing each button in the "Sensitivity runs" part, the Monte Carlo simulation with a specified number of iterations will be started (Figure 3-40). The outputs are shown on the "Sensitivity outputs" page. For example, Figure 3-41a and Figure 3-41b show the sensitivity analysis of the economic index of the Truck-Shovel and FIPCC systems respectively.

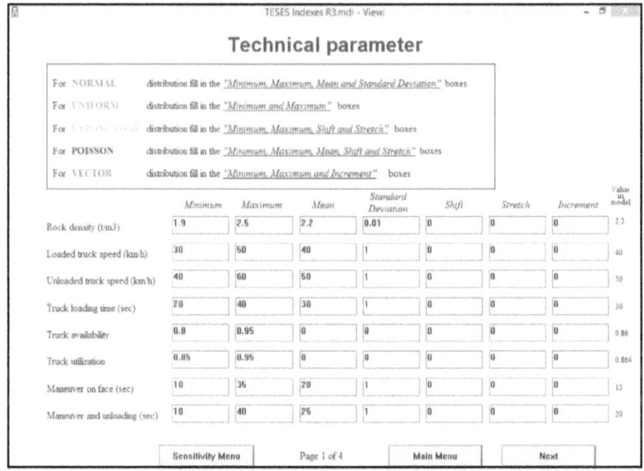

Figure 3-39. "Technical parameters" for the sensitivity analysis in TEcESaS Indexes software

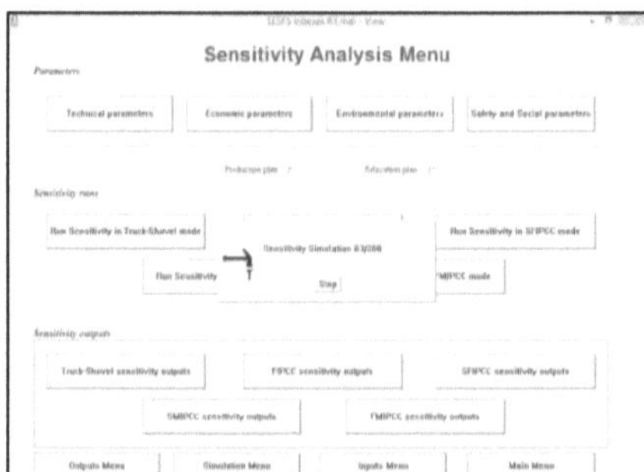

Figure 3-40. Sensitivity simulation by 200 iterations

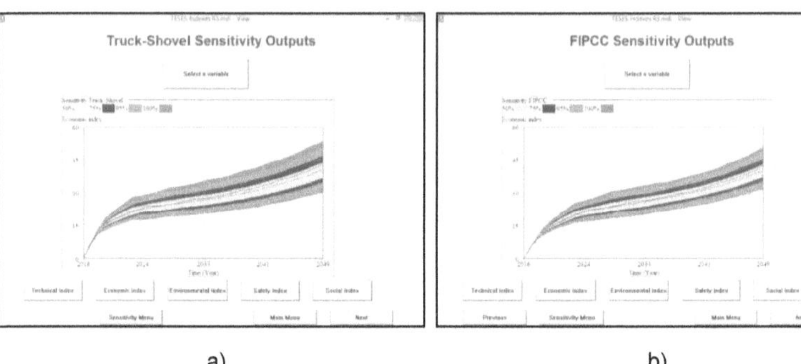

a)   b)

Figure 3-41. Sensitivity analysis output for the economic index in the a) Truck-Shovel and b) FIPCC systems

3.11.5.   Getting output for Sustainability Index software

This part is one of the most essential parts of the TEcESaS Indexes software because of getting the necessary outputs for running another software called "Sustainability Index" software. This software, which is thoroughly explained in Section 4.6, can be run just by importing the outputs of TEcESaS Indexes software as its inputs. Accordingly, this page must be run for all modes (Figure 3-42) if users want to continue the transportation system selection in Sustainability Index software. The outputs of this part will be saved as *.tab files that are readable for the Sustainability Index software.

Figure 3-42. "Getting output for Sustainability Software" in TEcESaS Indexes software

CHAPTER 4: Transportation system selection by analytical hierarchy process (AHP)

4.1. Developing the model

Analytical Hierarchy Process (AHP), one of the multi-criteria selection methods, is based on the measuring pairwise comparisons defined by the experts' judgment [113]. In this method, the main goal is to define the priorities among different quantities [114]. This method can be categorized in decision analysis, operation research, or both [114]. In the following sections[12], the steps of the AHP are explained through developing this study. Accordingly, interested readers are referred to the related references for additional information [114, 115].

The first step in the AHP analysis is determining the hierarchy for the making decision [116]. The hierarchy for this study, which is selecting the transportation system selection, can be defined as Figure 4-1.

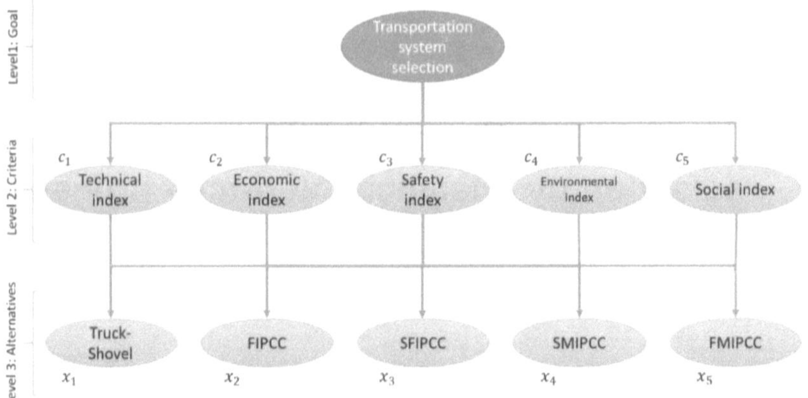

Figure 4-1. The hierarchy of the problem

As can be seen in Figure 4-1, the hierarch of the problem is defined in three different levels:

> Level 1: goal, which determines the outcome of the model.
> Level 2: criteria, which the decision is made based on them ($C = \{c_1, c_2, c_3, c_4, c_5\}$).
> Level 3: alternatives, which indicate the different transportation system that one of them should be selected ($X = \{x_1, x_2, x_3, x_4, x_5\}$).

Generally, decision-makers need to weigh each of the alternatives and, finally, select the one with the maximum value. Accordingly, a weight factor of $W = \{w_1, w_2, w_3, w_4, w_5\}$ will be generated in which w_i estimates the score of the alternative x_i [114]. Whenever w_j is bigger than w_i, the x_j is preferred to x_i.

---

[12] Sections 4.2 to 4.6 are published as the following paper:

Abbaspour, H. Drebenstedt, C. (2020). Introducing system dynamics and analytical hierarchy process based software for selecting the best transportation system in mines. *International Journal of Scientific & Technology Research* 9(3), 5648-5655.

4.2. Pairwise comparison matrix

In the pairwise process, it is possible to score the different alternatives concerning each other, which is more comfortable than getting each alternative an individual score. A pairwise comparison matrix, $A = (a_{ij})_{n \times n}$, is defined as follows:

$$A = \begin{pmatrix} a_{11} & a_{12} & \cdots & a_{1n} \\ a_{21} & a_{22} & \cdots & a_{2n} \\ \vdots & \vdots & \ddots & \vdots \\ a_{n1} & a_{n2} & \cdots & a_{nn} \end{pmatrix}$$

Equation 4-1

By substituting a_{ij} in matrix A with the related weights, based on Saaty's theory [117], the relation between two weights will be as follows:

$$a_{ij} \approx \frac{w_i}{w_j} \quad \forall i,j$$

Equation 4-2

$$A = \begin{pmatrix} \dfrac{w_1}{w_1} & \dfrac{w_1}{w_2} & \cdots & \dfrac{w_1}{w_n} \\ \dfrac{w_2}{w_1} & \dfrac{w_2}{w_2} & \cdots & \dfrac{w_2}{w_n} \\ \vdots & \vdots & \ddots & \vdots \\ \dfrac{w_n}{w_1} & \dfrac{w_n}{w_2} & \cdots & \dfrac{w_n}{w_n} \end{pmatrix}$$

Equation 4-3

To form the comparison matrix, relative weights can be scaled through numbers described in Table 4-1.

Table 4-1. Relative weights of the comparison matrix [118]

AHP Scale of Importance for comparison pair (a_{ij})	Numeric Rating	Reciprocal (decimal)
Extreme Importance	9	1/9 (0.111)
Very strong to extremely	8	1/8 (0.125)
Very strong Importance	7	1/7 (0.143)
Strongly to very strong	6	1/6 (0.167)
Strong Importance	5	1/5 (0.200)
Moderately to Strong	4	1/4 (0.250)
Moderate Importance	3	1/3 (0.333)
Equally to Moderately	2	1/2 (0.500)
Equal Importance	1	1 (1.000)

4.3. Priority vectors

Generally, a priority vector is a "numerical ranking of the alternatives that indicates an order of preference among them" [119]. Although there are different methods that a priority vector can be determined, the most important of them are:

4.3.1. Eigenvector method

In this method, the priority vector should be the principal eigenvector of the comparison matrix. If matrix A (Equation 4-3) is multiplied by weights (w), the following equation will result:

$$Aw = \begin{pmatrix} \dfrac{w_1}{w_1} & \dfrac{w_1}{w_2} & \cdots & \dfrac{w_1}{w_n} \\ \dfrac{w_2}{w_1} & \dfrac{w_2}{w_2} & \cdots & \dfrac{w_2}{w_n} \\ \vdots & \vdots & \ddots & \vdots \\ \dfrac{w_n}{w_1} & \dfrac{w_n}{w_2} & \cdots & \dfrac{w_n}{w_n} \end{pmatrix} \begin{pmatrix} w_1 \\ w_2 \\ \vdots \\ w_n \end{pmatrix} = \begin{pmatrix} nw_1 \\ nw_2 \\ \vdots \\ nw_n \end{pmatrix} = nw \qquad \text{Equation 4-4}$$

Vector w can be concluded from any pairwise comparison matrix A by solving the following equation:

$$\begin{cases} Aw = \lambda_{max}w \\ w^T 1 = 1 \end{cases} \qquad \text{Equation 4-5}$$

where λ_{max} is the maximum eigenvalue of A and $1 = (1, \dots, 1)^T$ [114].

4.3.2. Geometric mean method

Based on this method, each component of w is the result of the geometric mean of the elements on the respective row, which is divided by a normalized term [114]. Accordingly, w_i will be determined by the following equation:

$$p_i = \left(\prod_{j=1}^{n} a_{ij} \right)^{\frac{1}{n}} \Bigg/ \sum_{i=1}^{n} \left(\prod_{j=1}^{n} a_{ij} \right)^{\frac{1}{n}} \qquad \text{Equation 4-6}$$

4.3.3. Least square method

For obtaining the priority vector, the following optimization problem should be solved:

$$Minimize \quad \sum_{i=1}^{n} \sum_{j=1}^{n} \left(a_{ij} - \frac{w_i}{w_j} \right)^2 \qquad \text{Equation 4-7}$$

$$subject\ to \quad \sum_{i=1}^{n} w_i = 1, \quad w_i > 0 \ \forall_i$$

4.3.4. Normalized column method

In this method, all the columns should be normalized by dividing each component in the comparison matrix by the sum of the components in its column. The priority vector will be obtained by calculating the arithmetic mean of each row (Equation 4-8).

$$p_i = \sum_{i=1}^{n} \left(\frac{1}{n} \sum_{j=1}^{n} \frac{a_{ij}}{\sum_{k=1}^{n} a_{kj}} \right) \qquad \text{Equation 4-8}$$

4.4. Consistency index and consistency ratio

The consistency determines that "each direct comparison a_{ik} is exactly confirmed by all indirect comparisons $a_{ij}a_{jk} \; \forall j$" [114]. The consistency index (CI) proposed by Saaty [120] is as Equation 4-9.

$$CI(A) = \frac{\lambda_{max} - n}{n - 1} \qquad \text{Equation 4-9}$$

Consistency ratio (CR), which is a rescaled form of CI, obtains through dividing CI by a real number RI_n (random index) (Equation 4-10). RI_n is "an estimation of the average CI obtained from a large enough set of randomly generated matrices of size n" [114].

$$CR(A) = \frac{CI(A)}{RI_n} \qquad \text{Equation 4-10}$$

RI_n differs based on the matrix order, in which a matrix with higher order will have a bigger RI_n (Table 4-2).

Table 4-2. Values of RI_n [114]

n	3	4	5	6	7	8	9	10
RI_n	0.5247	0.8816	1.1086	1.2479	1.3417	1.4057	1.4499	1.4854

If the CR value is 0.1 or less, the comparison matrix is acceptable; otherwise, its components need to be revised.

4.5. Steps of transportation system selection by AHP

After introducing some important concepts that are widely used to define and solve the AHP, this section will present all the necessary transportation system selection steps.

4.5.1. Step 1: Pairwise comparison matrix

Referring to Section 4.2, the pairwise comparison matrix A will be as follows:

$$A = \begin{pmatrix} a_{11} & a_{12} & a_{13} & a_{14} & a_{15} \\ a_{21} & a_{22} & a_{23} & a_{24} & a_{25} \\ a_{31} & a_{32} & a_{33} & a_{34} & a_{35} \\ a_{41} & a_{42} & a_{43} & a_{44} & a_{45} \\ a_{51} & a_{52} & a_{53} & a_{54} & a_{55} \end{pmatrix} \qquad \text{Equation 4-11}$$

By substituting Equation 4-2 into Equation 4-11, matrix A will be constituted based on the relative weights (Equation 4-12 and Equation 4-13).

$$A = \left(\frac{w_i}{w_j}\right)_{5\times5} = \begin{pmatrix} \frac{w_1}{w_1} & \frac{w_1}{w_2} & \frac{w_1}{w_3} & \frac{w_1}{w_4} & \frac{w_1}{w_5} \\ \frac{w_2}{w_1} & \frac{w_2}{w_2} & \frac{w_2}{w_3} & \frac{w_2}{w_4} & \frac{w_2}{w_5} \\ \frac{w_3}{w_1} & \frac{w_3}{w_2} & \frac{w_3}{w_3} & \frac{w_3}{w_4} & \frac{w_3}{w_5} \\ \frac{w_4}{w_1} & \frac{w_4}{w_2} & \frac{w_4}{w_3} & \frac{w_4}{w_4} & \frac{w_4}{w_5} \\ \frac{w_5}{w_1} & \frac{w_5}{w_2} & \frac{w_5}{w_3} & \frac{w_5}{w_4} & \frac{w_5}{w_5} \end{pmatrix} \qquad \text{Equation 4-12}$$

$$A = \begin{pmatrix} 1 & a_{12} & a_{13} & a_{14} & a_{15} \\ \dfrac{1}{a_{12}} & 1 & a_{23} & a_{24} & a_{25} \\ \dfrac{1}{a_{13}} & \dfrac{1}{a_{23}} & 1 & a_{34} & a_{35} \\ \dfrac{1}{a_{14}} & \dfrac{1}{a_{24}} & \dfrac{1}{a_{34}} & 1 & a_{45} \\ \dfrac{1}{a_{15}} & \dfrac{1}{a_{25}} & \dfrac{1}{a_{35}} & \dfrac{1}{a_{45}} & 1 \end{pmatrix}$$

Equation 4-13

4.5.2. Step 2: Priorities (weights) for the criteria

In the second step, a priority matrix from the criteria, which are technical, economic, safety, environmental, and social indexes (c_1, c_2, c_3, c_4, c_5), should be constructed. The values of this matrix are based on the judgment of engineers and experts according to their experiences. The geometric mean method described in section 4.3.2 was applied for constructing the priority vector (Table 4-3). Through the programmed software, which will be explained in Section 4.6, the consistency index of this matrix will be calculated. The expert will be warned if the consistency index condition is not met (Section 4.4).

Table 4-3. Priority vector for the criteria

	c_1	c_2	c_3	c_4	c_5	$\left(\prod_{j=1}^{5} a_{ij}\right)^{\frac{1}{5}}$	p_i
c_1	1	a_{12}	a_{13}	a_{14}	a_{15}	$(1 \times a_{12} \times a_{13} \times a_{14} \times a_{15})^{\frac{1}{5}}$	$\left(\prod_{j=1}^{n} a_{1j}\right)^{\frac{1}{5}} \bigg/ \sum_{i=1}^{5}\left(\prod_{j=1}^{5} a_{ij}\right)^{\frac{1}{5}}$
c_2	a_{21}	1	a_{23}	a_{24}	a_{25}	$(a_{21} \times 1 \times a_{23} \times a_{24} \times a_{25})^{\frac{1}{5}}$	$\left(\prod_{j=1}^{n} a_{2j}\right)^{\frac{1}{5}} \bigg/ \sum_{i=1}^{5}\left(\prod_{j=1}^{5} a_{ij}\right)^{\frac{1}{5}}$
c_3	a_{31}	a_{32}	1	a_{34}	a_{35}	$(a_{31} \times a_{32} \times 1 \times a_{34} \times a_{35})^{\frac{1}{5}}$	$\left(\prod_{j=1}^{n} a_{3j}\right)^{\frac{1}{5}} \bigg/ \sum_{i=1}^{5}\left(\prod_{j=1}^{5} a_{ij}\right)^{\frac{1}{5}}$
c_4	a_{41}	a_{42}	a_{43}	1	a_{45}	$(a_{41} \times a_{42} \times a_{43} \times 1 \times a_{45})^{\frac{1}{5}}$	$\left(\prod_{j=1}^{n} a_{4j}\right)^{\frac{1}{5}} \bigg/ \sum_{i=1}^{5}\left(\prod_{j=1}^{5} a_{ij}\right)^{\frac{1}{5}}$
c_5	a_{51}	a_{52}	a_{53}	a_{54}	1	$(a_{51} \times a_{52} \times a_{53} \times a_{54} \times 1)^{\frac{1}{5}}$	$\left(\prod_{j=1}^{n} a_{5j}\right)^{\frac{1}{5}} \bigg/ \sum_{i=1}^{5}\left(\prod_{j=1}^{5} a_{ij}\right)^{\frac{1}{5}}$
Sum						$\sum_{i=1}^{5}\left(\prod_{j=1}^{5} a_{ij}\right)^{\frac{1}{5}}$	

4.5.3. Step 3: Local priorities (preferences) for the alternatives

This step determines which alternative is preferable concerning each criterion. It must be determined which one of the transportation systems is preferable based on the TEcESaS indexes. The content of this matrix (input) comes from the outputs resulted from the model in Vensim. The components of this matrix are resulted by comparably dividing their TEcESaS indexes. For instance, the components and the local priorities in the technical index for the different transportation systems are as Table 4-4.

In contrast with the priorities vector for criteria defined by the user or expert, the local priorities for alternatives result from the comparison of criteria indexes, which are output from the TEcESaS Indexes software. Hence, the consistency problem may happen, and need to find a way of resolving this issue. Accordingly, one of the most recent methods for resolving inconsistency in priorities matrix [121], which is based on reconstructing the pairwise comparison matrix by a corrected matrix W_s through the normalized priority vector w_n, was implemented (Equation 4-14):

$$W_s = w_n \cdot \left(\frac{1}{w_n}\right)^T \qquad\qquad \text{Equation 4-14}$$

By constituting all the local priorities for the different alternatives and criteria, the final matrix would be as Table 4-5.

Table 4-4. Local priority vector for the alternatives (technical index)

	x_1	x_2	x_3	x_4	x_5	$\left(\prod_{j=1}^{5} x_{ij}\right)^{\frac{1}{5}}$	$(lp_i)_{c_1}$
x_1	1	$\frac{T_1}{T_2}$	$\frac{T_1}{T_3}$	$\frac{T_1}{T_4}$	$\frac{T_1}{T_5}$	$(1 \times \frac{T_1}{T_2} \times \frac{T_1}{T_3} \times \frac{T_1}{T_4} \times \frac{T_1}{T_5})^{\frac{1}{5}}$	$\left(\prod_{j=1}^{n} x_{1j}\right)^{\frac{1}{5}} \Big/ \sum_{i=1}^{5}\left(\prod_{j=1}^{5} x_{ij}\right)^{\frac{1}{5}}$
x_2	$\frac{T_2}{T_1}$	1	$\frac{T_2}{T_3}$	$\frac{T_2}{T_4}$	$\frac{T_2}{T_5}$	$(\frac{T_2}{T_1} \times 1 \times \frac{T_2}{T_3} \times \frac{T_2}{T_4} \times \frac{T_2}{T_5})^{\frac{1}{5}}$	$\left(\prod_{j=1}^{n} x_{2j}\right)^{\frac{1}{5}} \Big/ \sum_{i=1}^{5}\left(\prod_{j=1}^{5} x_{ij}\right)^{\frac{1}{5}}$
x_3	$\frac{T_3}{T_1}$	$\frac{T_3}{T_2}$	1	$\frac{T_3}{T_4}$	$\frac{T_3}{T_5}$	$(\frac{T_3}{T_1} \times \frac{T_3}{T_2} \times 1 \times \frac{T_3}{T_4} \times \frac{T_3}{T_5})^{\frac{1}{5}}$	$\left(\prod_{j=1}^{n} x_{3j}\right)^{\frac{1}{5}} \Big/ \sum_{i=1}^{5}\left(\prod_{j=1}^{5} x_{ij}\right)^{\frac{1}{5}}$
x_4	$\frac{T_4}{T_1}$	$\frac{T_4}{T_2}$	$\frac{T_4}{T_3}$	1	$\frac{T_4}{T_5}$	$(\frac{T_4}{T_1} \times \frac{T_4}{T_2} \times \frac{T_4}{T_3} \times 1 \times \frac{T_4}{T_5})^{\frac{1}{5}}$	$\left(\prod_{j=1}^{n} x_{4j}\right)^{\frac{1}{5}} \Big/ \sum_{i=1}^{5}\left(\prod_{j=1}^{5} x_{ij}\right)^{\frac{1}{5}}$
x_5	$\frac{T_5}{T_1}$	$\frac{T_5}{T_2}$	$\frac{T_5}{T_3}$	$\frac{T_5}{T_4}$	1	$(\frac{T_5}{T_1} \times \frac{T_5}{T_2} \times \frac{T_5}{T_3} \times \frac{T_5}{T_4} \times 1)^{\frac{1}{5}}$	$\left(\prod_{j=1}^{n} x_{5j}\right)^{\frac{1}{5}} \Big/ \sum_{i=1}^{5}\left(\prod_{j=1}^{5} x_{ij}\right)^{\frac{1}{5}}$
Sum						$\sum_{i=1}^{5}\left(\prod_{j=1}^{5} x_{ij}\right)^{\frac{1}{5}}$	

T_i: Technical index of the alternative x_i

4.5.4. Step 4: Overall priorities (model synthesis)

In this step, an overall priority will be defined by considering both the priorities and local priorities. For calculating the overall priorities, both the priorities (weights) and local priorities (preferences) are merged into one table by inserting the priorities as the first row in the local priorities matrix (Table 4-6).

By multiplying the priorities in the local priorities and summing the components of each resulted row, the overall priorities will be calculated (Table 4-7).

Table 4-5. Local priorities matrix for different alternatives and criteria

	c_1	c_2	c_3	c_4	c_5
x_1	$(lp_1)_{c_1}$	$(lp_1)_{c_2}$	$(lp_1)_{c_3}$	$(lp_1)_{c_4}$	$(lp_1)_{c_5}$
x_2	$(lp_2)_{c_1}$	$(lp_2)_{c_2}$	$(lp_2)_{c_3}$	$(lp_2)_{c_4}$	$(lp_2)_{c_5}$
x_3	$(lp_3)_{c_1}$	$(lp_3)_{c_2}$	$(lp_3)_{c_3}$	$(lp_3)_{c_4}$	$(lp_3)_{c_5}$
x_4	$(lp_4)_{c_1}$	$(lp_4)_{c_2}$	$(lp_4)_{c_3}$	$(lp_4)_{c_4}$	$(lp_4)_{c_5}$
x_5	$(lp_5)_{c_1}$	$(lp_5)_{c_2}$	$(lp_5)_{c_3}$	$(lp_5)_{c_4}$	$(lp_5)_{c_5}$

Table 4-6. Priorities and local priorities

	c_1	c_2	c_3	c_4	c_5
	p_1	p_2	p_3	p_4	p_5
x_1	$(lp_1)_{c_1}$	$(lp_1)_{c_2}$	$(lp_1)_{c_3}$	$(lp_1)_{c_4}$	$(lp_1)_{c_5}$
x_2	$(lp_2)_{c_1}$	$(lp_2)_{c_2}$	$(lp_2)_{c_3}$	$(lp_2)_{c_4}$	$(lp_2)_{c_5}$
x_3	$(lp_3)_{c_1}$	$(lp_3)_{c_2}$	$(lp_3)_{c_3}$	$(lp_3)_{c_4}$	$(lp_3)_{c_5}$
x_4	$(lp_4)_{c_1}$	$(lp_4)_{c_2}$	$(lp_4)_{c_3}$	$(lp_4)_{c_4}$	$(lp_4)_{c_5}$
x_5	$(lp_5)_{c_1}$	$(lp_5)_{c_2}$	$(lp_5)_{c_3}$	$(lp_5)_{c_4}$	$(lp_5)_{c_5}$

Table 4-7. Overall priorities

	c_1	c_2	c_3	c_4	c_5	Overall priorities
x_1	$p_1 \cdot (lp_1)_{c_1}$	$p_2 \cdot (lp_1)_{c_2}$	$p_3 \cdot (lp_1)_{c_3}$	$p_4 \cdot (lp_1)_{c_4}$	$p_5 \cdot (lp_1)_{c_5}$	$\sum_{j=1}^{5} p_j \cdot (lp_1)_{c_j}$
x_2	$p_1 \cdot (lp_2)_{c_1}$	$p_2 \cdot (lp_2)_{c_2}$	$p_3 \cdot (lp_2)_{c_3}$	$p_4 \cdot (lp_2)_{c_4}$	$p_5 \cdot (lp_2)_{c_5}$	$\sum_{j=1}^{5} p_j \cdot (lp_2)_{c_j}$
x_3	$p_1 \cdot (lp_3)_{c_1}$	$p_2 \cdot (lp_3)_{c_2}$	$p_3 \cdot (lp_3)_{c_3}$	$p_4 \cdot (lp_3)_{c_4}$	$p_5 \cdot (lp_3)_{c_5}$	$\sum_{j=1}^{5} p_j \cdot (lp_3)_{c_j}$
x_4	$p_1 \cdot (lp_4)_{c_1}$	$p_2 \cdot (lp_4)_{c_2}$	$p_3 \cdot (lp_4)_{c_3}$	$p_4 \cdot (lp_4)_{c_4}$	$p_5 \cdot (lp_4)_{c_5}$	$\sum_{j=1}^{5} p_j \cdot (lp_4)_{c_j}$
x_5	$p_1 \cdot (lp_5)_{c_1}$	$p_2 \cdot (lp_5)_{c_2}$	$p_3 \cdot (lp_5)_{c_3}$	$p_4 \cdot (lp_5)_{c_4}$	$p_5 \cdot (lp_5)_{c_5}$	$\sum_{j=1}^{5} p_j \cdot (lp_5)_{c_j}$

4.6. Sustainability Index software[13]

The AHP method, which was explained in the previous sections, was developed in the Java programming language. Finally, a software called "Sustainability Index" software was created that enables users to easily handle the process of transportation system selection (Appendix III). Generally, this software has three steps:

1. Importing data from "TEcESaS Indexes software"
2. Entering the pairwise comparison of the different indexes
3. Calculating the sustainability index for the transportation system alternatives

---

[13] This software was published in the Mendeley Dataset with the DOI 10.17632/kxkcmvdgw7.2

4.6.1. Importing data from "TEcESaS indexes application"

In this step, the output resulted from the TEcESaS Indexes software, which is described in Section 3.11.5, is imported. This page in the software is shown in Figure 4-2.

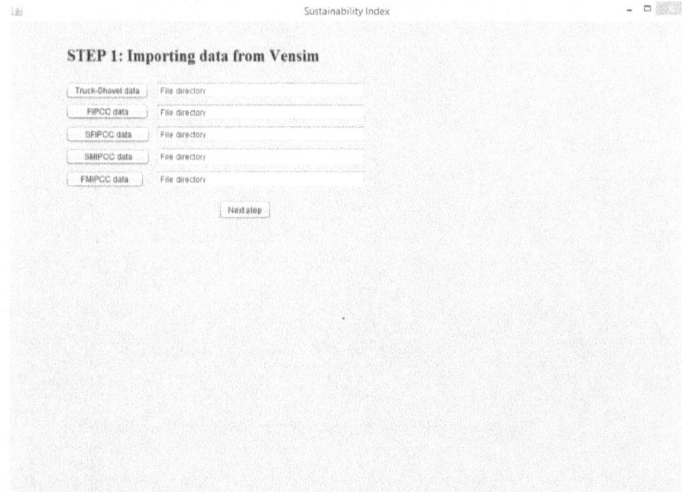

Figure 4-2. Importing data from TEcESaS indexes software

By pressing each button on this page, the user can import the output file, a *.tab file from the TEcESaS Indexes software (Figure 4-3a). If the user does not import all the necessary files, an error will appear to show that they need to enter all the files (Figure 4-3b).

4.6.2. Entering the pairwise comparison of the different indexes

In this step, all the pairwise comparisons, which were thoroughly explained in Section 4.5.1, should be entered. It can be handled by a single expert as well as multiple experts (group decision making). In the single-mode, the expert just needs to fill the upper side of the matrix, and the lower side will be filled automatically (Figure 4-4). If the pairwise comparisons are not selected completely or selected so that the consistency ratio is bigger than 0.1, error messages will be appeared (Figure 4-5).

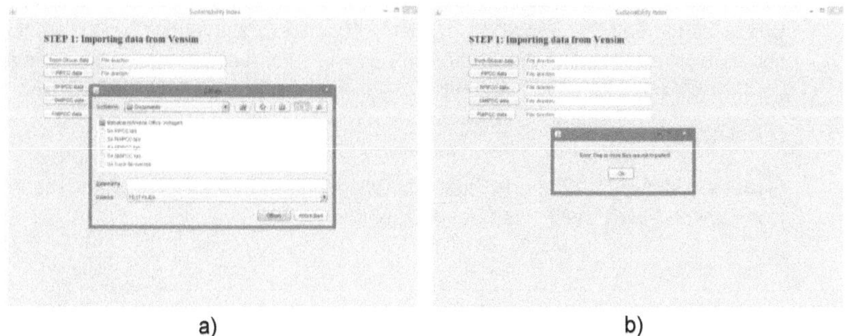

a)   b)

Figure 4-3. Importing data a) from the specific file direction b) not importing a direction error

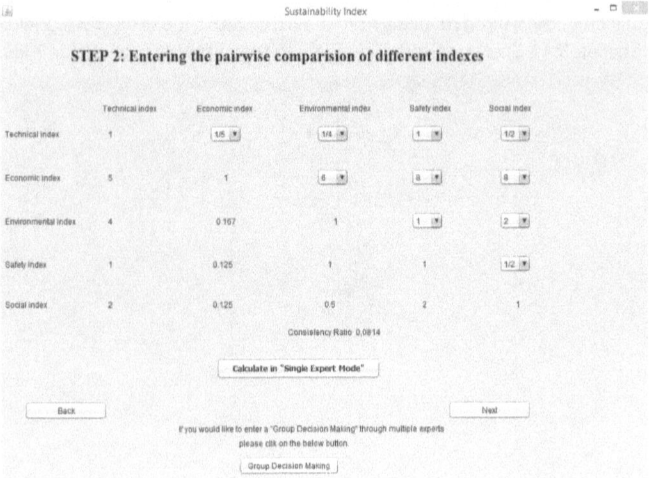

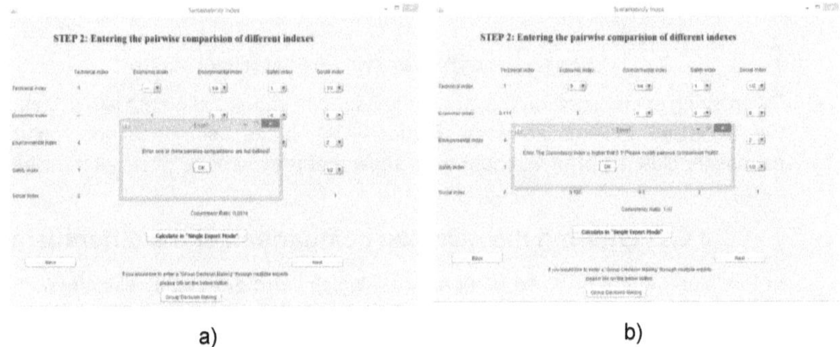

Figure 4-4. Pairwise comparisons in Sustainability index software

a)   b)

Figure 4-5. Error message for a) not completing the pairwise comparison and b) a consistency ratio bigger than 0.1

In the "Group Decision Making" mode, it is asked from each expert merely to specify four comparisons of "Technical / Economic", "Economic / Environmental", "Environmental / Safety", and "Safety / Social". It is mainly because of preventing inconsistency in the comparison matrix, and the other comparison items will be automatically calculated [116]. For instance, the "Technical / Environmental" pairwise comparison will be calculated as "$\dfrac{Technical}{Economic} \times \dfrac{Economic}{Environmental}$".

Figure 4-6 shows the "Experts Evaluation Matrix" filled by four experts and the "Pairwise Comparison Matrix", which is calculated based on their evaluation.

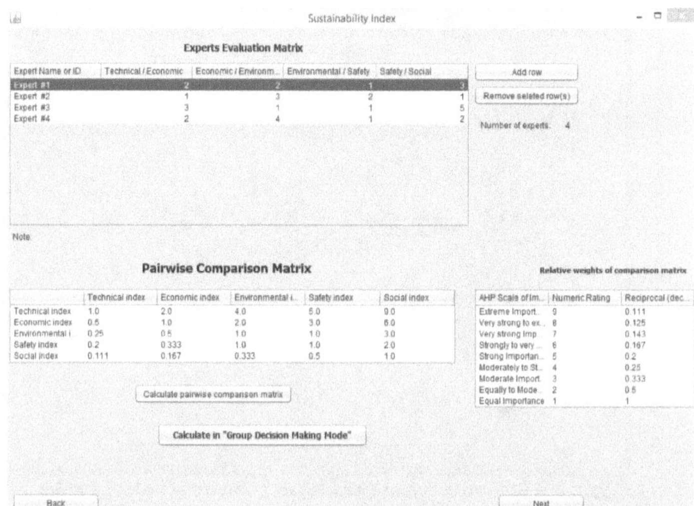

Figure 4-6. "Group Decision Making" page of the Sustainability Index software

4.6.3. Calculating the sustainability index for transportation systems

After completing the previous steps, by pressing the "Calculate" button in the calculation page, the software starts calculating the sustainability index of each transportation system based on the TEcESaS indexes. Finally, it will list the selected transportation system that should be used in each year of the project. Figure 4-7 and Figure 4-8 show a sample result of a single expert mode in this step in the tabular and graphical format, respectively.

Year	Truck-Shovel	FIPCC	SFIPCC	SMIPCC	FMIPCC	Best Sustainability Index	Selected System
2016.0	0.252	0.208	0.193	0.172	0.175	0.252	Truck-Shovel
2017.0	0.252	0.206	0.194	0.175	0.181	0.252	Truck-Shovel
2018.0	0.248	0.206	0.195	0.178	0.182	0.248	Truck-Shovel
2019.0	0.243	0.206	0.196	0.181	0.181	0.243	Truck-Shovel
2020.0	0.232	0.203	0.195	0.178	0.196	0.232	Truck-Shovel
2021.0	0.226	0.201	0.195	0.177	0.205	0.226	Truck-Shovel
2022.0	0.222	0.201	0.191	0.176	0.212	0.222	Truck-Shovel
2023.0	0.220	0.201	0.189	0.175	0.218	0.220	Truck-Shovel
2024.0	0.219	0.201	0.188	0.174	0.221	0.221	FMIPCC
2025.0	0.216	0.201	0.187	0.173	0.218	0.218	FMIPCC
2026.0	0.212	0.199	0.184	0.177	0.225	0.225	FMIPCC
2027.0	0.209	0.197	0.182	0.179	0.229	0.229	FMIPCC
2028.0	0.207	0.196	0.181	0.181	0.232	0.232	FMIPCC
2029.0	0.207	0.195	0.180	0.179	0.235	0.235	FMIPCC
2030.0	0.208	0.194	0.180	0.179	0.237	0.237	FMIPCC

Figure 4-7. Calculating the sustainability indexes of the different transportation systems

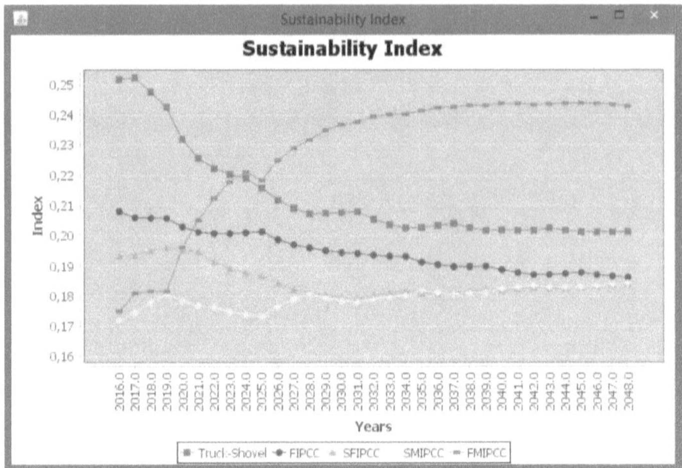

Figure 4-8. Graphical format of the sustainability indexes of the transportation systems

CHAPTER 5: Case study, results and discussion

5.1. Specifications of the hypothetical copper mine

In this section, all the copper mine's general specifications, including technical, economic, environmental safety, and social parameters, are described. They are all illustrated in Table 5-1 to Table 5-4. These parameters are used as inputs in the TEcESaS indexes application to simulate the model.

In Table 5-1, the technical specifications are merely considered for ore. It is assumed that the transportation system selection is going to be performed for ore reserve. All the parameters mentioned as "in the default mode" need to be filled if there is no specific production plan. Accordingly, the model will be simulated through these quantities.

In Table 5-2, the copper price is based on the average yearly price between January 1996 to December 2015 [122]. The copper price trend is shown in Figure 5-1.

The weight factors assigned in the case study are presented in Table 5-5.

Table 5-1. Technical specifications of the hypothetical copper mine

Parameter	Quantity	Unit	Reference
Mine and mill specification			
Ore reserve	700,000,000	t	
Start year of ore extraction	2016		
Number of holidays in a year	15	days	
Number of shifts	3		
Working hours per shift	8	h	
Rock density	2.2	t/m^3	
Average ore grade (in default mode)	0.3	%	
Road's grade	10	%	
Road's width	30	m	
Number of faces	3		
Surface elevation (in default mode)	2100	m	
Pit's bottom elevation (in default mode)	1800	m	
Mill recovery	80	%	
Combined smelter/refinery recovery	95	%	
Truck specification			
Truck capacity	60	m^3	[123]
Loaded truck speed	30	km/h	
Unloaded truck speed	40	km/h	
Truck loading time	30	sec	
Truck availability	86	%	[27]
Truck utilization	86.4	%	[27]
Maneuver on face	15	sec	
Maneuver and unloading	20	sec	
Delays	10	sec	
Engine load factor[14] (loaded truck)	35	%	[124]
Engine load factor (unloaded truck)	25	%	[124]
Unloaded truck mass	165	t	[123]
Rolling resistance	5	%	
Shovel specification			
Number of shovels per face	1		
Swell factor	30	%	
Load cycle time	20	sec	

---

[14] The portion of full power required by truck [124]

Propel time factor	90	%	
Shovel availability	86	%	[27]
Shovel utilization	81	%	[27]
IPCC and spreader/stacker specification			
FIPCC availability	85	%	[27]
FIPCC utilization	85	%	[27]
SFIPCC availability	85	%	[27]
SFIPCC utilization	85	%	[27]
SMIPCC availability	83.7	%	[27]
SMIPCC utilization	87.8	%	[27]
FMIPCC availability	84	%	[27]
FMIPCC utilization	83.8	%	[27]
Spreader/Stacker availability	87	%	[27]
Spreader/Stacker utilization	91.7	%	[27]
IPCCs relocation (in default mode)			
FIPCC interval years of relocation	15		
SFIPCC interval years of relocation	10		
SMIPCC interval years of relocation	5		
FIPCC depth of relocation	100	m	
SFIPCC depth of relocation	75	m	
SMPCC depth of relocation	50	m	
Conveyor belt specification			
Surcharge angle	20	°	
Conveyor belt inclination	16	°	
Ambient temperature correction factor	3		[67]
Carrying run factor	0.018		[67]
Fractional resistance of plows	0	lbs	
Resistance of trippers and stackers	0	lbs	
Resistance of belt-cleaning devices	0	lbs	
Skirtboard length	10	ft	
Skirtboard friction factor	0.276		[67]
Depth of material touching skirtboard	10	ft	
Idler spacing	3	in	
Idler diameter	6	in	
Length of each conveyor set	150	m	
Conveyor belt availability	92.2	%	[27]
Conveyor belt utilization	89.9	%	[27]

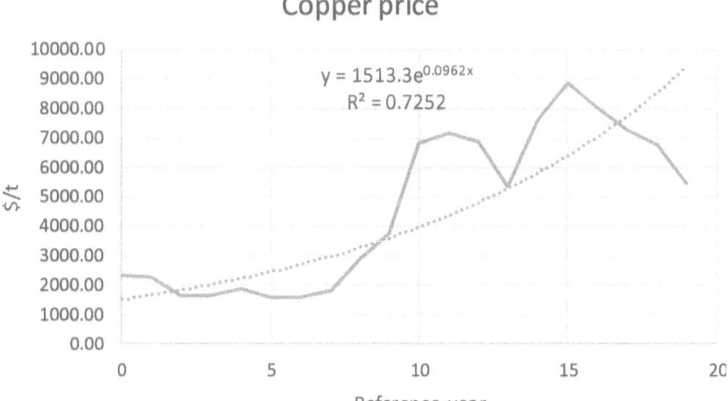

Figure 5-1. The copper price trend between 1996 and 2015 (the reference year 0 and 19 respectively)

Table 5-2. Economic specifications of the hypothetical copper mine

Parameter	Quantity	Unit
Truck specification		
Truck price (base year)	550,000	$
Truck lifespan	10	years
Fuel price (base year)	1	$/L
Tire life	6,000	h
Number of tires per truck	6	
Hourly wage of a truck driver (base year)	20	$/h
Conveyor belt specification		
Conveyor belt price (base year)	100	$/m
Conveyor belt lifespan	5	years
Conveyor set price (base year)	750,000	$
Conveyor set lifespan	15	years
Electricity price (base year)	0.07	$/kWh
Workforce per 1000 m conveyor belt	2	
Hourly wage of conveyor labor (base year)	20	$/h
IPCC specification		
Crusher station purchase price (base year)	60	M$
FIPCC purchase price (base year)	80	M$
SFIPCC purchase price (base year)	90	M$
SMIPCC purchase price (base year)	110	M$
FMIPCC purchase price (base year)	150	M$
Relocation cost in depth (base year)	1000	$/m
Rates, factors, and copper price		
Inflation rate	2	%
Wage annual increment rate	1.5	%
Repair factor	0.5	
Depreciation rate	13	%
Copper price (Exponential trend)	$1513e^{0.096x}$	$/t
Base year of copper price	1996	

Table 5-3. Environmental specifications of the hypothetical copper mine

Parameter	Quantity	Unit
Emissions		
Heat rate	10059	Btu/kWh
Heat content	19.28	MBtu/kWh
CO_2 emission factor (electricity)	0.36	
NO_x emission factor (fuel)	0.034	
SO_2 coefficient	30	
CO_2 oxidization factor	0.99	
SO_2 oxidization factor	0.98	
Carbon content of the fuel	86	%
Sulfur content of the fuel	0.0015	%
Weight percent of sulfur in fuel	3	%
Fuel mass	840	gr/L
Particulate matters		
Emission factor $PM_{2.5}$ (conveyor belt)	6.5×10^{-6}	gr/t

Emission factor PM_{10} (conveyor belt)	2.3×10^{-5}	gr/t
Emission factor PM_{30} (conveyor belt)	7×10^{-5}	gr/t
Surface material silt content	0.2	%
Water consumption		
Average dust control efficiency	75	%
Average hourly evaporation rate	0.28	mm/h
Time interval of spraying water	0.5	hour

Table 5-4. Safety and social specifications of the hypothetical copper mine

Parameter	Quantity	Unit
Injury rate per accident	2	
Hours of training per lost-time injury	5	hours
Hours of training per person	20	hours
Death rate per accident	0.5	
Accident rate per 1000 trucks in hour	1	

Table 5-5. Weight factors for the different items in the defined indexes

Item	symbol	Weight factor
Technical		
System availability	w(SA)	0.3
System utilization	w(SU)	0.3
Conveyor belt power consumption	w(CBP)	0.3
Loaded truck power consumption	w(LTP)	0.05
Unloaded truck power consumption	w(UTP)	0.05
Economic		
Income	w(I)	0.333
Total operating costs	w(TOC)	0.333
Total capital costs	w(TCC)	0.333
Environmental		
Particulate matter rate	w(PMR)	0.2
Emission rate	w(ER)	0.3
Equivalent noise level	w(ENL)	0.2
Water consumption rate	w(WCR)	0.3
Safety		
Traffic density	w(TD)	0.05
Traffic flow	w(TF)	0.05
Average truck speed	w(ATS)	0.1
Lost time injury frequency rate	w(LTIFR)	0.2
Number of conveyor sets	w(NCS)	0.1
Training	w(T)	0.5
Social		
Health index	w(HI)	0.3
Number of employees	w(NE)	0.2
Training	w(T)	0.5

For running the model, the production and relocation plan for all the transportation systems are considered as Table 5-6. Although the production plan of the transportation systems could differ from one to another [18, 125, 126], in this case, the identical production plan is considered for all of the transportation systems for the sake of comparability. Nevertheless, the TEcESaS application has this ability to accept any kind of production plan as its input through modify the Excel file "ProductionRelocation.xlsx", attached to the TEcESaS application.

Table 5-6. Production and relocation plan for the hypothetical copper mine

| year | Ore production (t) | Average ore grade (%) | Average faces elevation (m) | Crusher / IPCC elevation (m) | | | | | ADFR[1] (m) | SE[2] (m) |
				Outside pit crusher	FIPCC	SFIPCC	SMIPCC	FMIPCC		
2016	21,500,000	0.3	2100	2100	2100	2100	2100	2100	500	2100
2017	21,500,000	0.3	2080	2100	2100	2100	2100	2080	400	
2018	21,500,000	0.29	2080	2100	2100	2100	2100	2080	300	
2019	21,500,000	0.4	2060	2100	2100	2100	2100	2060	200	
2020	21,500,000	0.29	2060	2100	2100	2100	2100	2060	700	
2021	21,500,000	0.35	2060	2100	2100	2100	2100	2060	600	
2022	21,500,000	0.34	2040	2100	2100	2100	2040	2040	500	
2023	21,500,000	0.37	2040	2100	2100	2100	2040	2040	400	
2024	21,500,000	0.38	2020	2100	2100	2100	2040	2020	300	
2025	21,500,000	0.35	2000	2100	2100	2100	2040	2000	1000	
2026	21,500,000	0.34	2000	2100	2100	2100	2040	2000	950	
2027	21,500,000	0.36	1980	2100	2100	2100	2000	1980	900	
2028	21,500,000	0.37	1980	2100	2000	2000	2000	1980	850	
2029	21,500,000	0.4	1980	2100	2000	2000	2000	1980	800	
2030	21,500,000	0.45	1980	2100	2000	2000	2000	1980	750	
2031	21,500,000	0.5	1960	2100	2000	2000	2000	1960	700	
2032	21,500,000	0.55	1960	2100	2000	2000	2000	1960	1200	
2033	21,500,000	0.55	1940	2100	2000	2000	1960	1940	1100	
2034	21,500,000	0.55	1920	2100	2000	2000	1960	1920	1000	
2035	21,500,000	0.6	1900	2100	2000	2000	1960	1900	950	
2036	21,500,000	0.5	1880	2100	2000	2000	1960	1880	1300	
2037	21,500,000	0.51	1880	2100	2000	2000	1960	1880	1200	
2038	21,500,000	0.52	1860	2100	2000	2000	1960	1860	1100	
2039	21,500,000	0.49	1840	2100	2000	2000	1960	1840	1000	
2040	21,500,000	0.48	1820	2100	2000	1900	1900	1820	900	
2041	21,500,000	0.47	1820	2100	2000	1900	1900	1820	850	
2042	21,500,000	0.45	1820	2100	2000	1900	1900	1820	1400	
2043	21,500,000	0.43	1800	2100	2000	1900	1900	1800	1300	
2044	21,500,000	0.42	1800	2100	2000	1900	1860	1800	1200	
2045	21,500,000	0.39	1800	2100	2000	1900	1860	1800	1100	
2046	21,500,000	0.4	1780	2100	2000	1900	1860	1780	1000	
2047	21,500,000	0.4	1780	2100	2000	1900	1860	1780	900	
2048	12,000,000	0.41	1780	2100	2000	1900	1860	1780	800	
Total	700,000,000									

1. Average distance from the face to the ramp
2. Surface elevation

5.2. Technical parameters and index

In the following subsections, the run model results for some technical parameters and the technical index will be discussed.

5.2.1. System availability

As shown in Figure 5-2, Truck-Shovel and FMIPCC systems have the highest and lowest system availability, respectively. This can be interpreted as different attitudes of these systems in truck usage, which is the highest in the former and is the lowest in the latter. On the other hand, in the Truck-Shovel system, the hybrid type provides more availability than a series system because of the parallel type in the hybrid. As previously mentioned, the parallel type gives more availability for the system mainly because of the capability to run the whole system even if one or more components are not available. Other systems (FIPCC, SFIPCC, and SMIPCC), which are a combination of trucks and conveyor belts, have a system availability ranging between the Truck-Shovel and FMIPCC systems. However, the SMIPCC system represents the lower system availability than the FIPCC and SFIPCC systems due to the lower quantity of trucks.

5.2.2. System utilization

In contrast with the system availability, the Truck-Shovel system shows lower system utilization than the FMIPCC system due to more trucks in this system and, consequently, more stoppage time for trucks. However, by increasing the number of trucks, the average system utilization will be increased (Figure 5-3). The FIPCC, SFIPCC, and SMIPCC systems show a relatively close system utilization percentage; however, in some intervals, one represents more system utilization than two others.

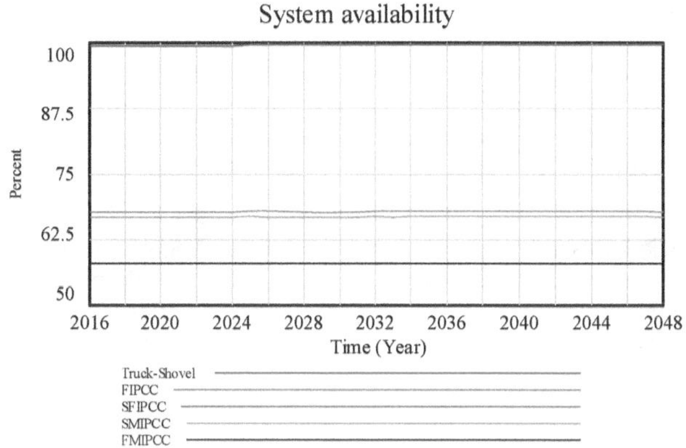

Figure 5-2. System availability for the different transportation systems

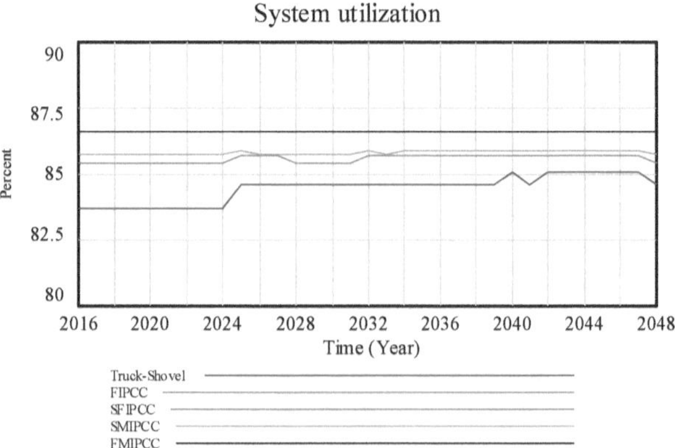

Figure 5-3. System utilization for the different transportation systems

5.2.3. Power consumption

The power consumption is the sum of the three parameters of the conveyor belt power, the loaded truck power, and the unloaded truck power, which are explained in the following.

5.2.3.1.   Conveyor belt power

Figure 5-4 shows the conveyor power consumption in the different transportation systems. Truck-Shovel system results in null conveyor belt power due to not having any conveyor belt. On the contrary, the conveyor belt power in the FMIPCC system will be gradually increased by the increment in the conveyor belt length. The conveyor belt power is dependent on the conveyor length, which is affected by the relocation of the IPCC systems through the mine life. For instance, the FIPCC system in this case study is relocated just once in 2028 into the lower elevation. Accordingly, the total length of the conveyor belt will be increased. This scenario will be varied for the SFIPCC and the SMIPCC systems because of the different relocation plan.

The conveyor belt power in the FMIPCC system is continuously increasing because of the continuous increment in conveyor length by progressing the face (Figure 5-4). However, for the other systems, it differs based on the relocation plan of the relevant system. For instance, the SMIPCC system goes deeper than the SFIPCC system, has more conveyor belt length, and finally, more conveyor belt power will result.

5.2.3.2.   Loaded truck power

Unlike the Truck-Shovel system, which consumes the highest amount of power for transporting ore, the FMIPCC system has no power consumption led by trucks obviously because of not using trucks in this system (Figure 5-5). In addition, by increasing the route's length for delivering ore from the faces to the crusher, more power of trucks will be needed. Each system that applies more trucks in operation consumes more power as well. After the Truck-Shovel system, three other systems of FIPCC, SFIPCC, and SMIPCC are placed in the next ranks, respectively.

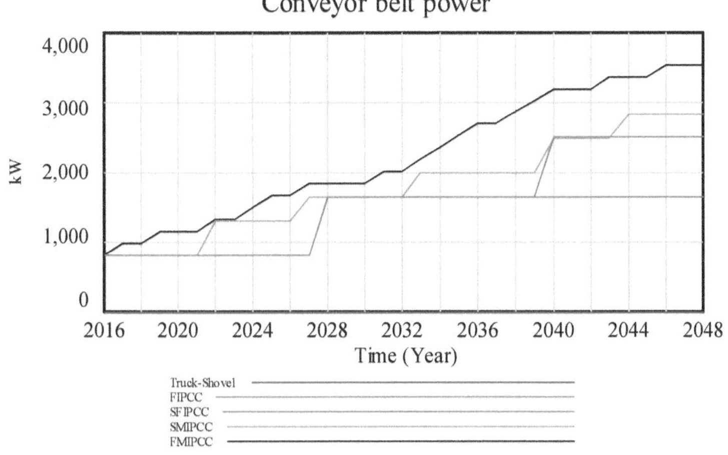

Figure 5-4. Conveyor belt power consumption for the different transportation

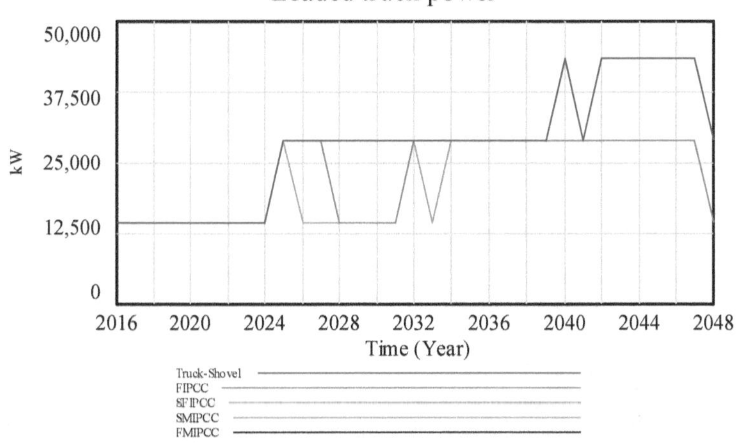

Figure 5-5. Loaded truck power for the different transportation systems

5.2.3.3. Unloaded truck power

Since the unloaded trucks in this case study pass a route from the top to the bottom of the pit, the road grade will be minus in Equation 3-13. As a result, the unloaded truck power will be negative (Figure 5-6). As it is evident in this figure, the lowest unloaded truck power belongs to the Truck-Shovel system.

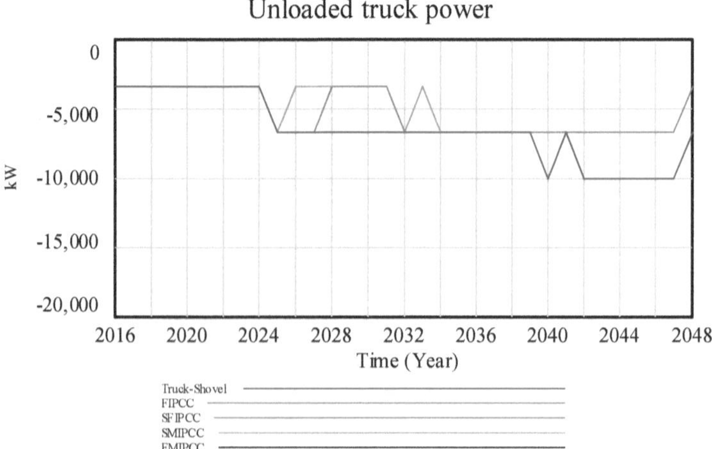

Figure 5-6. Unloaded truck power for the different transportation systems

5.2.4. Other technical parameters

In this section, the results of the simulation for some other technical parameters will be presented.

5.2.4.1.   Number of trucks

Through the mine's life, the number of trucks will be increased (Figure 5-7) because the length of delivering ore will be increased while the yearly ore production should be fulfilled. The Truck-Shovel system with the highest and the FMIPCC system with no trucks stands at the first and last rank, respectively.

5.2.4.2.   Hourly truck production and truck cycle time

Truck production per hour will be decreased by increasing the length of the route (Figure 5-8), mainly due to the increment of the truck cycle time (Figure 5-9). Accordingly, each truck needs to deliver a fixed amount of ore, equivalent to its capacity, to the crusher while the route length is increased. However, in the FIPCC, SFIPCC, and SMIPCC systems, trucks' hourly production will be improved by relocating the IPCC systems to a closer distance to the faces.

5.2.4.3.   Total truck traveling distance and total conveyor length

In the Truck-Shovel system, the total truck traveling distance will be continuously increased; however, in the FIPCC, SFIPCC, and SMIPCC, this parameter will be adjusted (Figure 5-10) after the relocation. On the contrary, the FMIPCC system has the highest length of the conveyor belt than the other types of IPCC systems (Figure 5-11).

5.2.4.4.   IIPCC capacity

IPCC capacity is calculated based on the maximum production per year through the mine's life. Additionally, the IPCC capacity is dependent on the IPCC availability and utilization, which vary for the IPCC systems (Table 5-7).

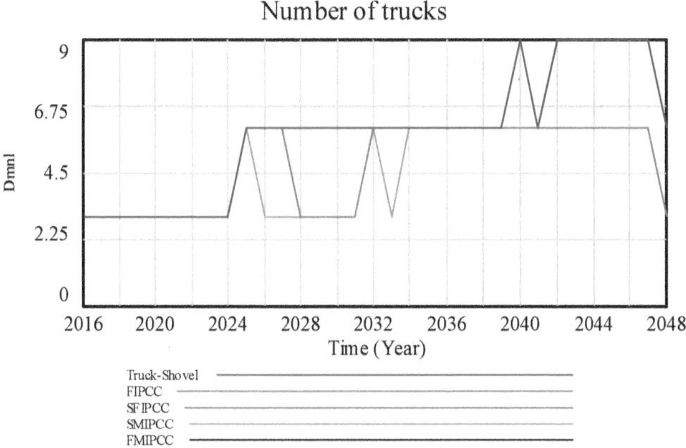

Figure 5-7. *Number of trucks for the different transportation systems*

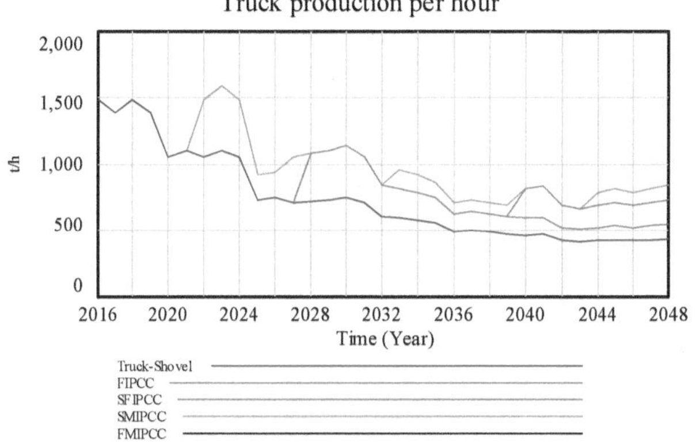

Figure 5-8. *Truck production per hour for the different transportation systems*

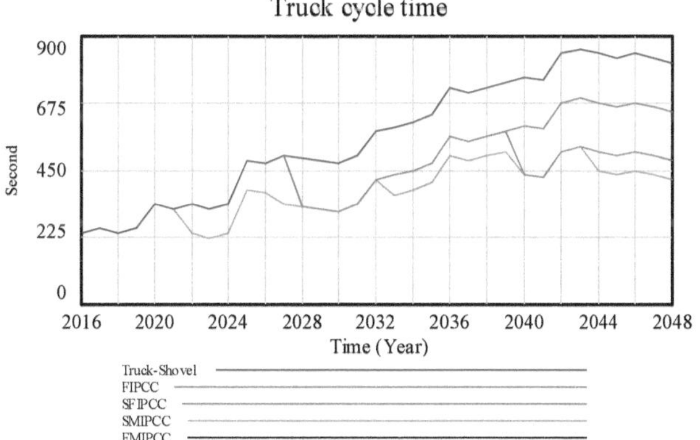

Figure 5-9. *Truck cycle time for the different transportation systems*

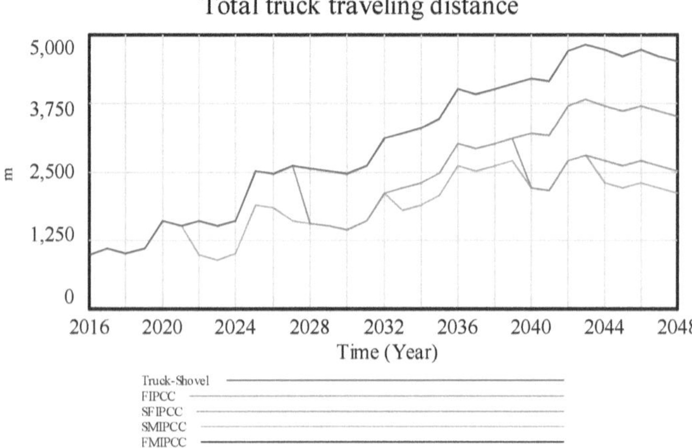

Figure 5-10. Total truck traveling distance for the different transportation systems

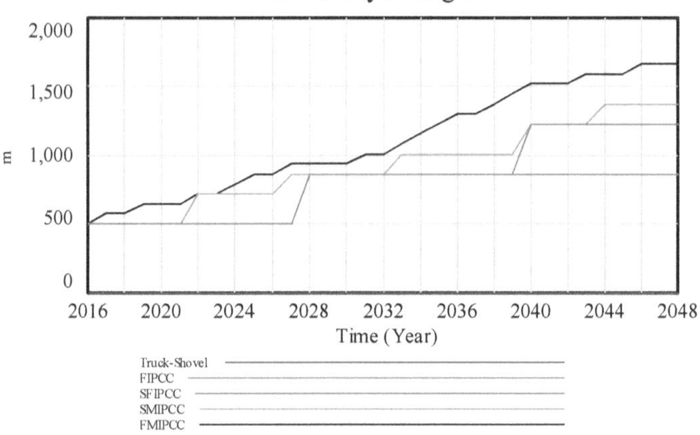

Figure 5-11. Total conveyor length for the different transportation systems

Table 5-7. IPCC systems capacity

System	FIPCC	SFIPCC	SMIPCC	FMIPCC
Capacity (t/h)	3543	3543	3483	3636

5.2.5. Technical index

The technical index for the transportation system alternatives is shown in Figure 5-12. As it can be seen in this figure, the highest technical index goes to the FMIPCC system. It can be interpreted as a much lower power consumption than the other systems (Figure 5-4). However, the technical index trend is decreasing due to the increment in power consumption through the years. The Truck-Shovel system stands as the second-highest technical index for most years of the project. Nevertheless, the FIPCC, SFIPCC, and SMIPCC systems have the occasionally higher technical index in some years, e.g., in 2028.

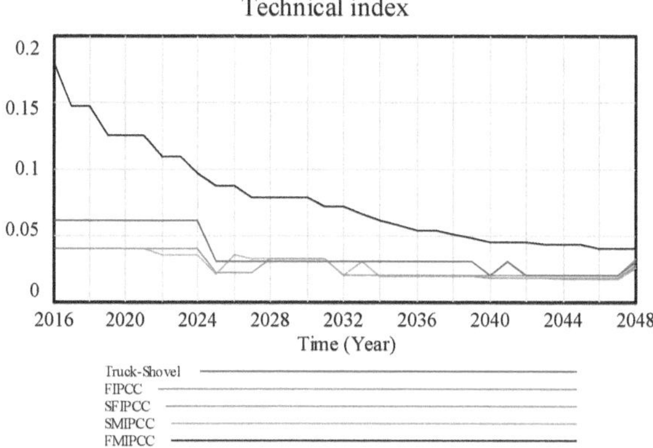

Figure 5-12. Technical index for the different transportation systems

5.3. Economic parameters and index

The simulation results for some economic parameters, and the economic index will be presented in the following subsections.

5.3.1. Truck operating costs

Any transportation system with more trucks will result in higher truck operating costs. Accordingly, the Truck-Shovel system will be at first, and the FMIPCC with no truck will be at the last place in the truck operating costs item (Figure 5-13).

5.3.2. Truck capital costs

As it is shown in Figure 5-14, the truck capital costs happen periodically in some years of the project. It is because of the increment in the number of trucks by progressing the project and purchasing new trucks. Besides, at the end of the truck's lifespan, buying a new truck and substituting it with the old one is foreseen.

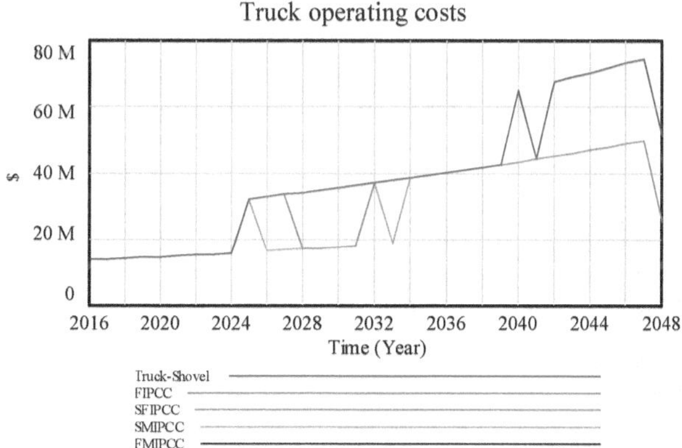

Figure 5-13. Truck operating costs for the different transportation systems

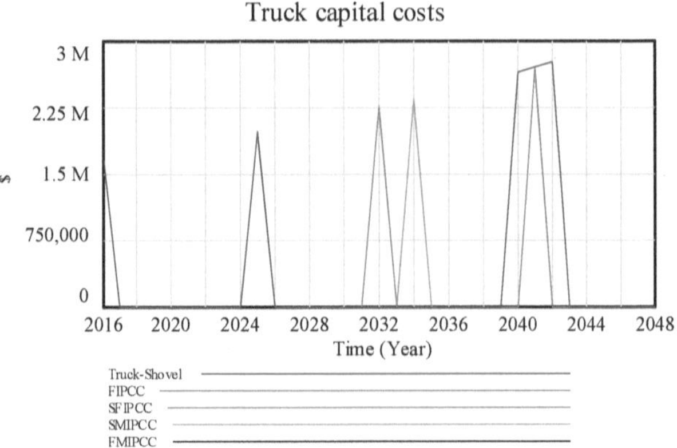

Figure 5-14. Truck capital costs for the different transportation systems

5.3.3. Conveyor belt operating costs

The FMIPCC system with the longest conveyor belt shows the highest conveyor belt operating costs, and the Truck-Shovel system with no conveyor belt results in zero conveyor belt costs (Figure 5-15). In the second, third, and fourth ranks, the SMIPCC, SFPCC, and FIPCC systems are placed. However, the FIPCC and SFIPCC show the same conveyor belt operating costs until 2039, in which the relocation of the SFIPCC takes place this year (Figure 5-15).

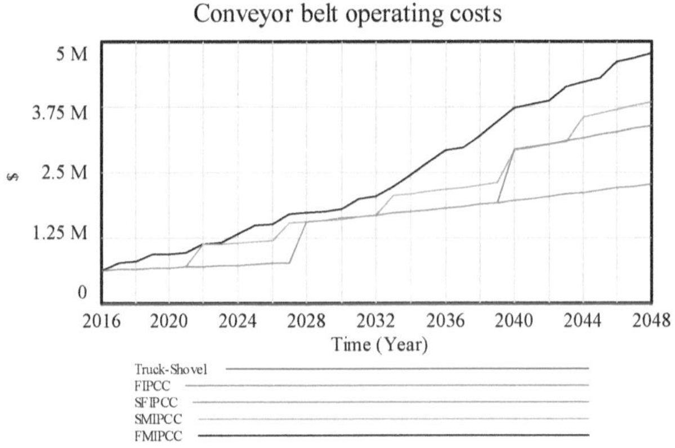

Figure 5-15. Conveyor belt operating costs for the different transportation systems

5.3.4. Conveyor belt capital costs

Conveyor belt capital costs include the increment of the conveyor belt length and the replacement cost of the conveyor belt after its lifespan. Hence, the FMIPCC system experiences a continuous and Truck-Shovel system will have no conveyor capital costs (Figure 5-16).

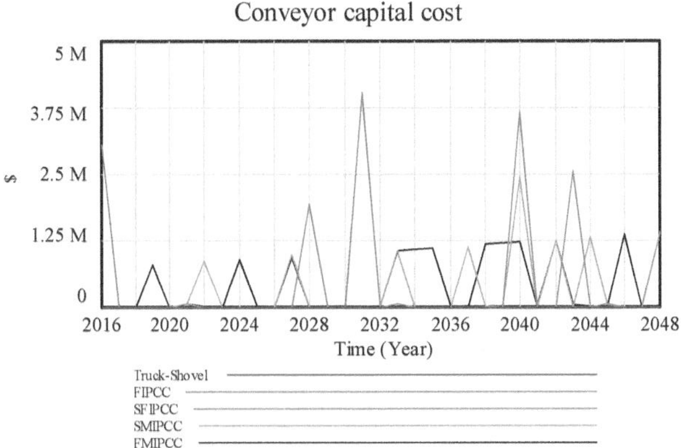

Figure 5-16. Conveyor belt capital costs for the different transportation systems

5.3.5. IPCC relocation cost

The IPCC relocation cost, which is a function of the relocation depth and the inflation rate, occurs continuously for the FMIPCC system because of its consecutive relocation through progressing the face. However, the relocation cost for the FIPCC system just happens in 2028, which is relocated once through the project. This cost is imposed on the project in 2040 in the SFIPCC and 2022, 2027, 2033, 2040, and 2044 in the SMIPCC system (Figure 5-17).

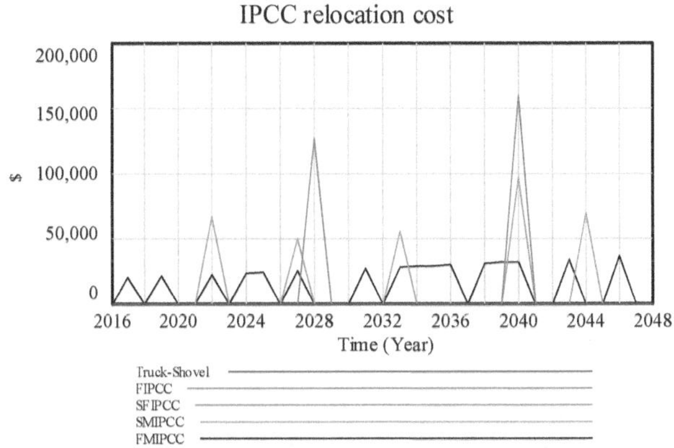

Figure 5-17. IPCC relocation cost for the different transportation systems

5.3.6. Total operating costs

Total operating costs, which is the sum of the conveyor belt operating costs, truck operating costs, and IPCC relocation costs along with the mine's life, are represented in Figure 5-18. While the FMIPCC and SMIPCC systems show the lowest and the second-lowest total operating cost through the years, respectively, the other transportation systems represent different quantities. Generally, it can be said that after 2028, the total operating costs of the SMIPCC system are lower than the FIPCC and SFIPCC systems.

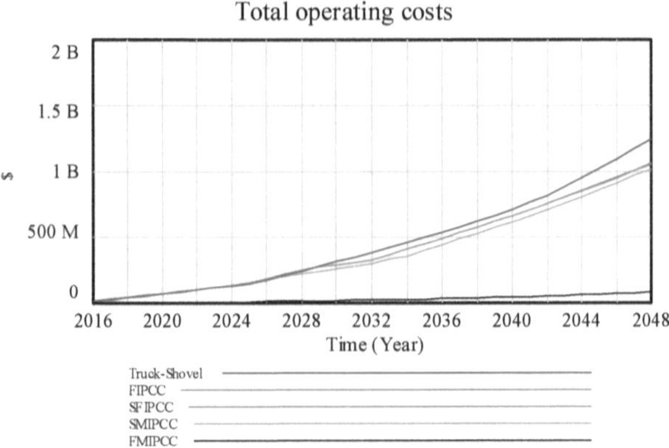

Figure 5-18. Total operating costs for the different transportation systems

5.3.7. Total capital costs

As it was expected and is shown in Figure 5-19, the highest total capital costs belong to the FMIPCC system. It is mainly due to the high investment cost of purchasing this system. In the next ranks of the highest total capital costs are placed the SMIPCC, SFIPCC, FIPCC, and Truck-Shovel systems, respectively.

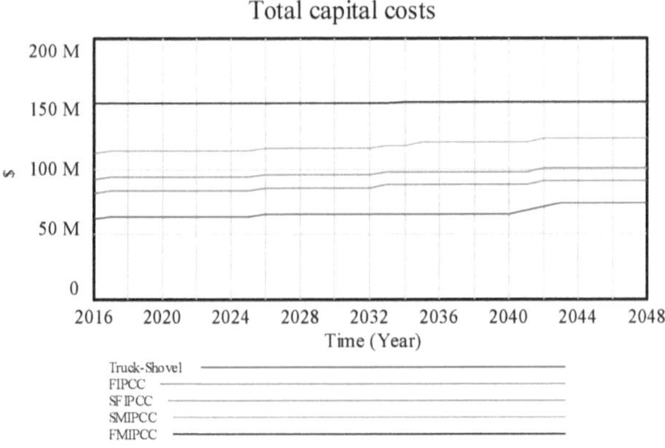

Figure 5-19. Total capital costs for the different transportation systems

5.3.8. Other economic parameters

Some other economic parameters are mentioned in the following.

5.3.8.1. Fuel and electricity cost

Fuel and electricity as energy sources for trucks and conveyor belts are vital to measuring. Figure 5-20 and Figure 5-21 show the cost of fuel and electricity for the different transportation systems. As expected, their trend is increasing due to the increment in the number of trucks and the conveyor belt's length through the project.

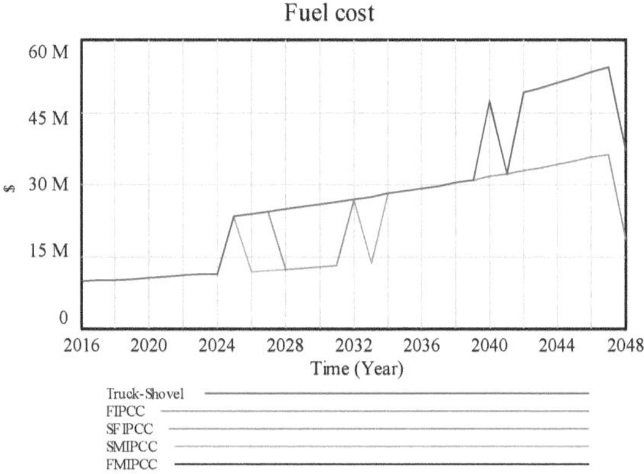

Figure 5-20. Fuel cost for the different transportation systems

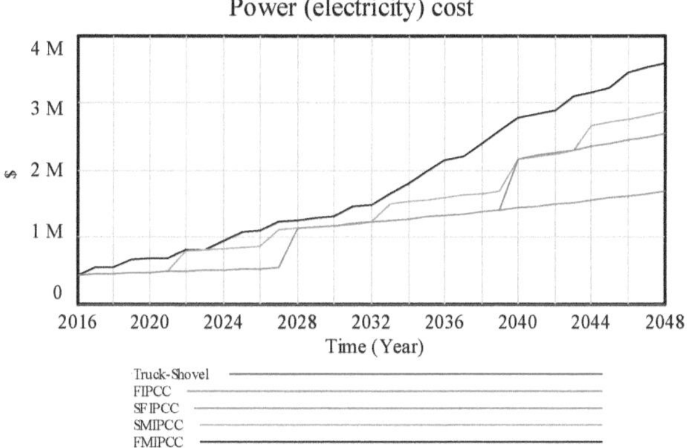

Figure 5-21. Electricity cost for the different transportation systems

5.3.8.2.   Truck and conveyor belt labor costs

Labors that work in the trucks and the conveyor belts as drivers and inspectors are imposing a major part of the project's operating costs. Generally, truck drivers' labor cost is much higher than the labor cost on the conveyor belts. These costs are shown in Figure 5-23 and Figure 5-23 for the truck and the conveyor belt labor costs.

5.3.9. Economic index

Unlike the higher economic index in the Truck-Shovel system from the start of the project to 2021, the FMIPCC showed the highest economic index afterward (Figure 5-24). It can be interpreted that although the capital cost of the FMIPCC system is the highest at the start of the project, its operating costs are the lowest throughout the project. Furthermore, the increment rate in the capital and the operating costs for this system is much lower than the others.

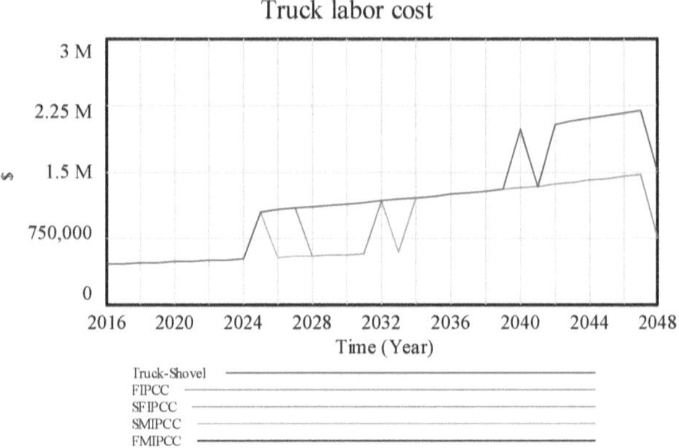

Figure 5-22. Truck labor costs for different transportation systems

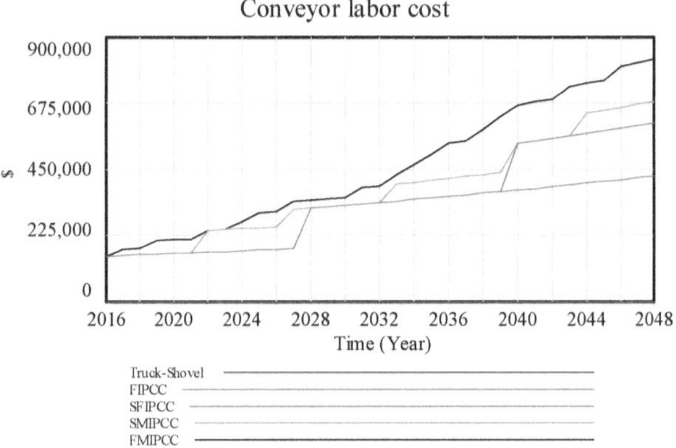

Figure 5-23 Conveyor belt labor cost for the different transportation systems

5.4. Environmental parameters and index

Some of the outcomes for the environmental parameters are presented in the following subsections.

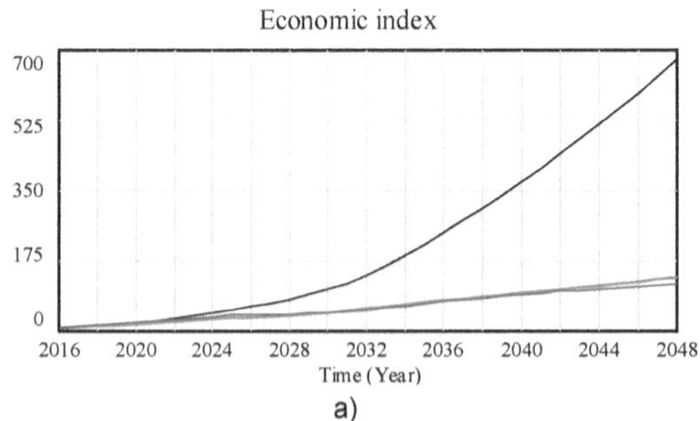

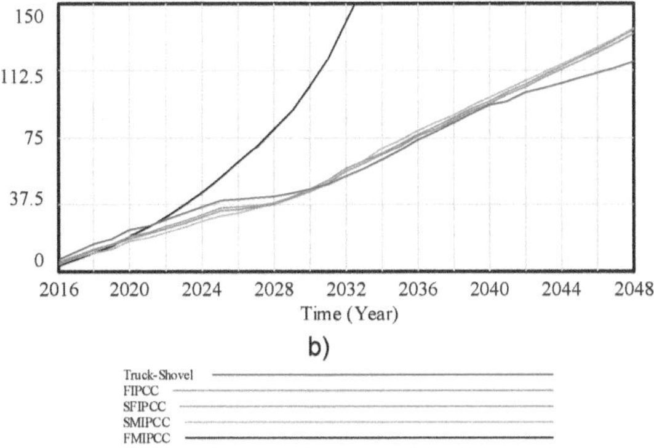

Figure 5-24. Economic index for different transportation systems a) general view and b) zoomed-in view

5.4.1. Total emissions

The total emissions, which are the sum of the emissions from trucks and conveyor belts, are shown in Figure 5-25. As can be seen in this figure, until 2029, the FIPCC system produces the highest emissions, while afterward, the Truck-Shovel system is the transportation system with the highest total emission. The FMIPCC system produces the lowest amount of emissions during the mine's life.

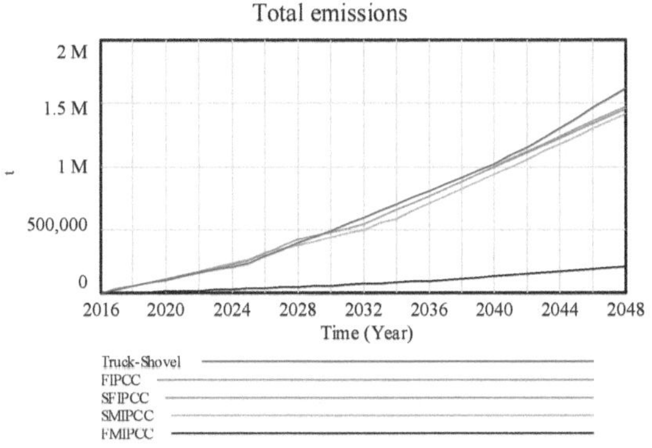

Figure 5-25. Total emissions for the different transportation systems

5.4.2. Total particulate matters

The total particulate matters, which are the sum of the particulate matters stemming from trucks and conveyor belts, are shown in Figure 5-26. Due to the lack of trucks in the FMIPCC system, the total particulate matters produced by this system are much lower than the others. It shows the significant contribution of trucks in producing particulate matter.

5.4.3. Total water consumption

The FMIPCC system consumes the lowest amount of water throughout the project (Figure 5-27). It can be interpreted mainly because of operating just conveyor belts in this system, while in the others, trucks are operating as well. It clearly will increase the total water consumption. Except for this transportation system, the order of the other transportation systems varies in the total consuming water. However, the Truck-Shovel system stands as the second-lowest total water consumer in the project.

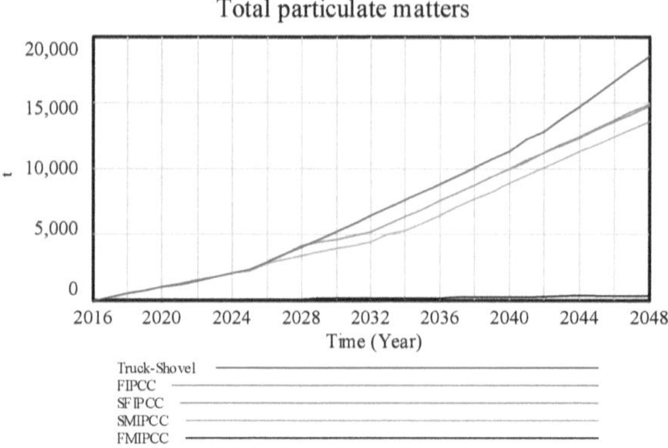

Figure 5-26. Total particulate matters for the different transportation systems

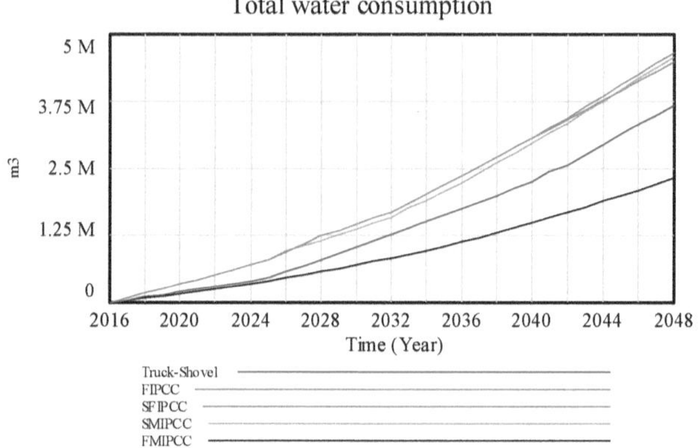

Figure 5-27. Total water consumption for the different transportation systems

5.4.4. Equivalent noise level

The FMIPCC represents the highest noise level among all the transportation systems (Figure 5-28). It can be explained as the more number of idlers in the conveying system than the FIPCC, SFIPCC, and SMIPCC systems. The Truck-Shovel with not having a conveying system and producing lower noise in trucks rather than conveyor belts shows the lowest equivalent noise level.

5.4.5. Environmental index

While the FMIPCC shows the best environmental index (Figure 5-29), the environmental index of other transportation systems varies in different time intervals.

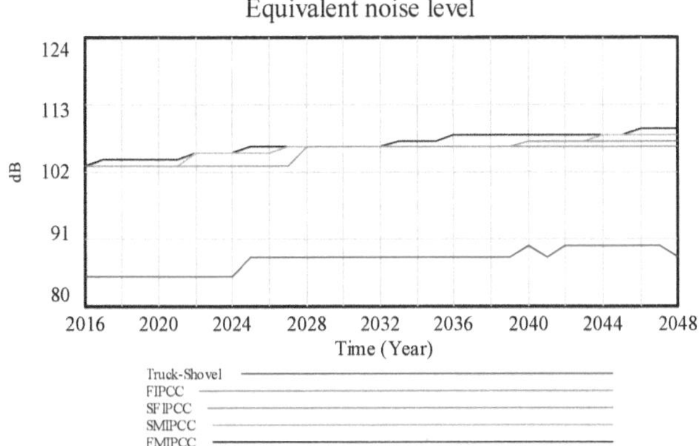

Figure 5-28. Equivalent noise level for the different transportation

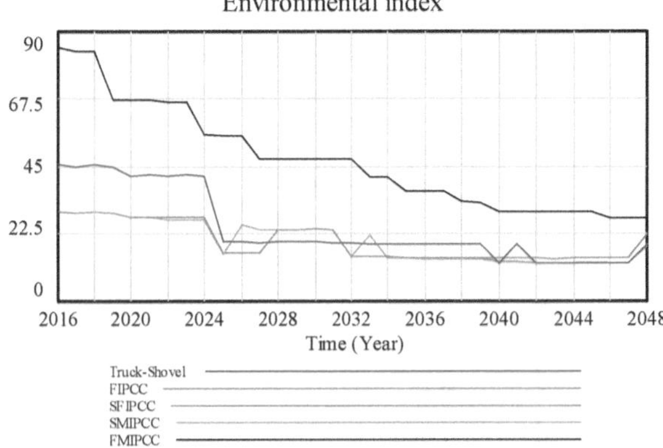

Figure 5-29. Environmental index for the different transportation systems

5.5. Safety parameters and index

Some of the safety parameters and the safety index will be explained in the following sections.

5.5.1. Traffic flow and traffic density

As it can be seen in Figure 5-30 and Figure 5-31, both the traffic flow and the traffic density for the Truck-Shovel, FIPCC, and SFIPCC systems have a decreasing trend. It is because of increasing the total truck traveling distance, the number of trucks, and the truck cycle time. On the contrary, the traffic flow for the SMIPCC system is increasing, and its traffic density shows a relatively constant trend. It depicts that despite fewer trucks in this system, the total traveling distance and the truck cycle time are not high enough to reduce the traffic flow and the traffic density. In addition, the traffic flow and the traffic density in the FMIPCC system are zero because no truck is operating in this system.

5.5.2. Lost time injury frequency rate

The results showed that the lost time injury frequency rate in the FMIPCC system is the lowest (Figure 5-32). It can be interpreted based on the lowest number of employees and not using trucks in this system, which leads to lower fatality and injury rates. In contrast, other systems show different lost time injury frequency rates, which are relatively close together. Additionally, its trend is decreasing, which can be interpreted as the increment in the training per lost-time injury.

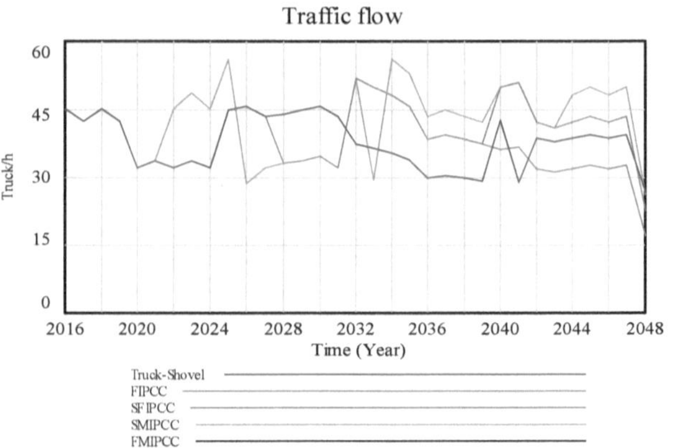

Figure 5-30. Traffic flow for the different transportation systems

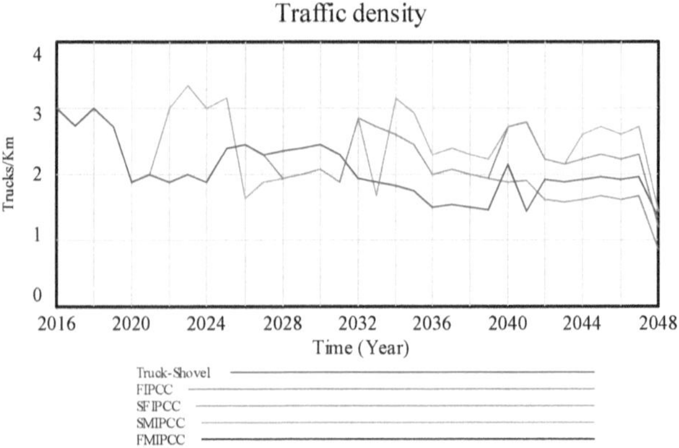

Figure 5-31. Traffic density for the different transportation systems

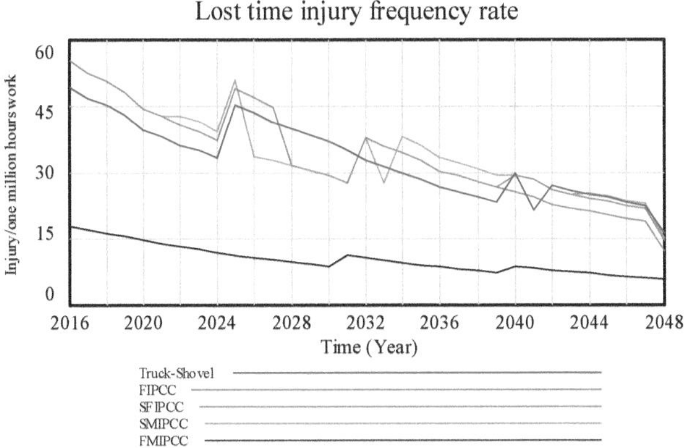

Figure 5-32. Lost time injury frequency rate for the different transportation systems

5.5.3. Safety index

While the best safety index belongs to the FMIPCC system, the lowest safety index goes to the SMIPCC system (Figure 5-33). It can be explained as working just conveyor belts in the FMIPCC system, while in the SMIPCC, trucks are operating likewise.

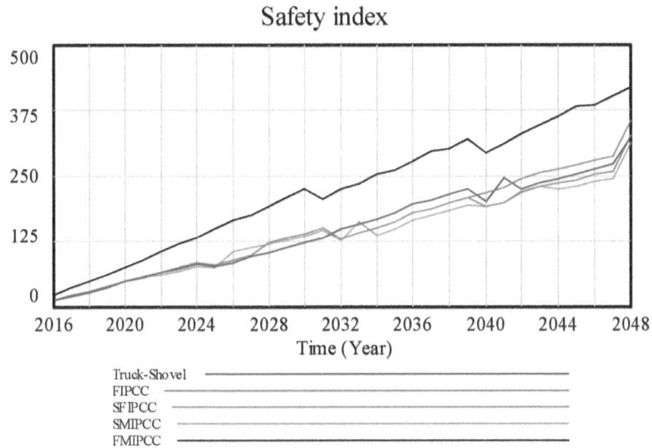

Figure 5-33. Safety index for the different transportation systems

5.6. Social parameters and index

The outcomes of the two most important social parameters, which are training and employment, will be discussed in the following sub-sections.

5.6.1. Training

The hours of training, mostly the safety training given to the employees, are the highest in the Truck-Shovel system (Figure 5-34). It is because of more employees in this system, which naturally need more hours of training. Additionally, the higher lost time injury frequency rate demands more training hours. However, its difference with the

FIPCC, SFIPCC, and SMIPCC systems is not considerable. On the contrary, the FMIPCC system experiences a big difference in hours of training. The higher training factor results in a higher social index.

5.6.2. Number of employees

Since the number of truck drivers is higher than the conveyor belt operators, the number of employees in the Truck-Shovel system is the highest consequently; however, its difference with the FIPCC, SFIPCC, and SMIPCC systems is not notable (Figure 5-35). On the other hand, the FMIPCC system stands at the lowest rank due to operating only through the conveyor belt.

5.6.3. Health index

As shown in Figure 5-36, the FMIPCC system shows the highest health index compared to the other transportation system alternatives. It results from less production of the particulate matter, emission and the noise level, and the highest safety index.

5.6.4. Social index

Since the training and the number of employees, which are the variables for determining social index, are higher in the Truck-Shovel and FIPCC systems, they stood in the first place in terms of the social index (Figure 5-37). In contrast, the FMIPCC placed in the last rank due to the lowest hours of training and employees.

5.7. Sensitivity analysis

There is a part of the TEcESaS Index software that sensitivity analysis can be handled. It can be done by considering one or more constants in the model as distributions, which five distributions of Normal, Uniform, Exponential, Poisson, and Vector are designed in this software. As was explained in Section 3.11.4, the sensitivity analysis is based on the Monte Carlo simulation by generating random numbers in the specified distribution. In this software, four different confident intervals of 50%, 75%, 95%, and 100% are applied. These intervals indicate how many percentages of the results are placed between two upper and lower limits. For instance, if 100 outputs out of 200 sensitivity runs are placed between 1 and 2, the confident intervals of 50% for the range [1,2] will be defined. This process and ranges will be automatically handled and determined by TEcESaS Indexes software. These confidence intervals are portrayed with yellow, blue, green, and gray colors for 50%, 75%, 95%, and 100%, respectively. The red and blue lines in the sensitivity graph show the mean values in the run and its default value.

As an example, a sensitivity analysis was performed, and the constants of Table 5-8 with specified distributions were taken into account. This sensitivity analysis for indexes is shown in Figure 5-38 to Figure 5-42 for different transportation systems.

From these figures, it can be generally concluded that the Truck-Shovel and the FMIPCC systems have the highest and lowest sensitivity respectively to the changes in the related constants. One of the reasons could be the changes in trucks' cycle time (e.g., the truck loading time, maneuver on the face, delays, etc.), which affect the Truck-Shovel system.

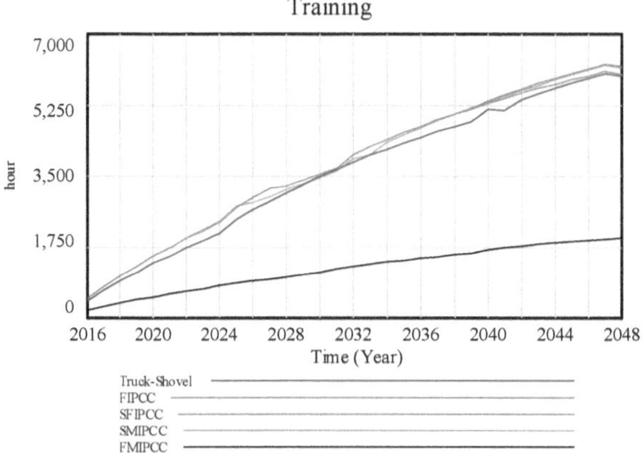

Figure 5-34. Training for different transportation systems

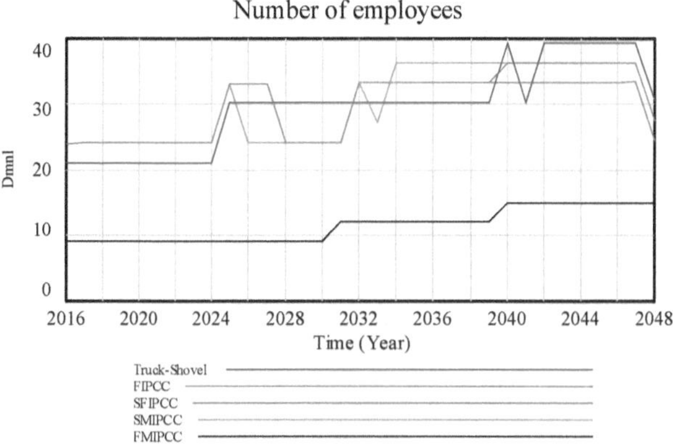

Figure 5-35. Number of employees for different transportation systems

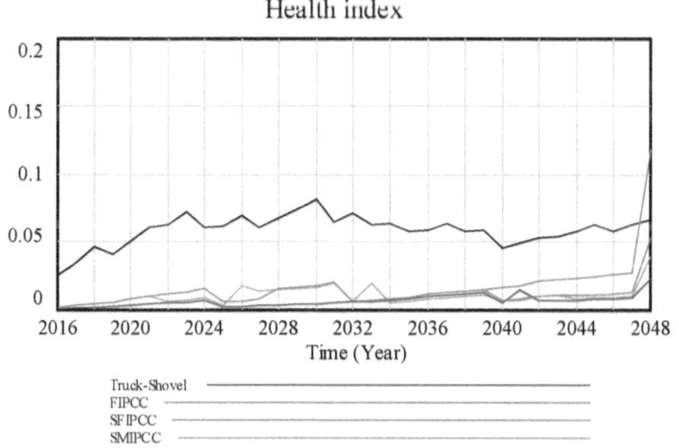

Figure 5-36. Health index for different transportation systems

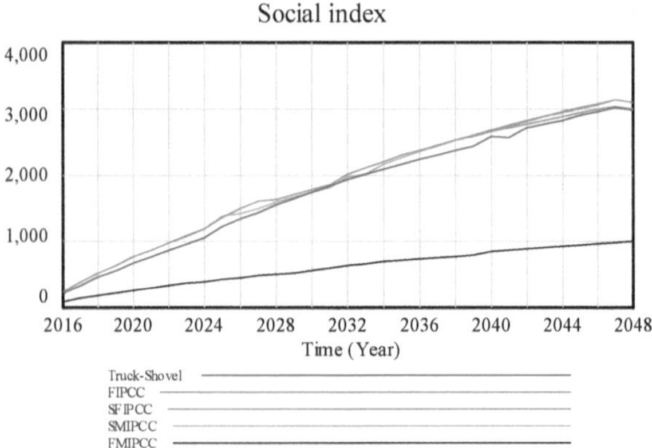

Figure 5-37. Social index for the different transportation systems

Table 5-8. Constants and their distribution in the sensitivity analysis

Constant	Unit	Distribution	Min	Max	Mean	Standard deviation
Technical parameters						
Rock density	t/m³	Normal	1.9	2.5	2.2	0.01
Average ore grade	%	Normal	0.2	0.6	0.35	0.01
Loaded truck speed	km/h	Normal	30	50	40	1
Unloaded truck speed	km/h	Normal	40	60	50	1
Truck loading time	s	Normal	20	40	30	1
Truck availability	%	Uniform	0.8	0.95		
Truck utilization	%	Uniform	0.85	0.95		
Maneuver on face	s	Normal	10	35	20	1
Maneuver and unloading	s	Normal	10	40	25	1
Delays	s	Normal	5	40	20	1
Shovel availability	%	Uniform	0.8	0.95		
Shovel utilization	%	Uniform	0.8	0.95		
Load cycle time (shovel)	s	Normal	15	25	20	1
FIPCC availability	%	Uniform	0.85	0.95		
FIPCC utilization	%	Uniform	0.85	0.95		
SFIPCC availability	%	Uniform	0.85	0.97		
SFIPCC utilization	%	Uniform	0.85	0.97		
SMIPCC availability	%	Uniform	0.8	0.95		
SMIPCC utilization	%	Uniform	0.8	0.95		
FMIPCC availability	%	Uniform	0.8	0.95		
FMIPCC utilization	%	Uniform	0.85	0.95		
Conveyor belt availability	%	Uniform	0.85	0.95		
Conveyor belt utilization	%	Uniform	0.8	0.95		
Conveyor belt inclination	°	Uniform	8	16		
Economic parameters						
Truck price	$	Normal	500,000	650,000	550,000	100
Truck lifespan	year	Normal	23	27	25	1
Tire life	hour	Normal	3500	4500	4000	100
Conveyor belt price	$/m	Normal	80	150	110	10
Conveyor belt lifespan	year	Normal	4	6	5	0.5
Relocation cost in depth	$/m	Normal	800	2000	1500	100
Environmental parameters						
Heat rate	Btu/kWh	Normal	9500	11000	10000	100
Dust control efficiency	%	Normal	75	95	80	5
Hourly evaporation rate	mm/h	Normal	0.25	0.35	0.3	0.1

Safety and social parameters				
Injury rate per accident	person	Uniform	1	3
Death rate per accident	person	Uniform	0	2
Hours of training per person	hour	Uniform	15	35

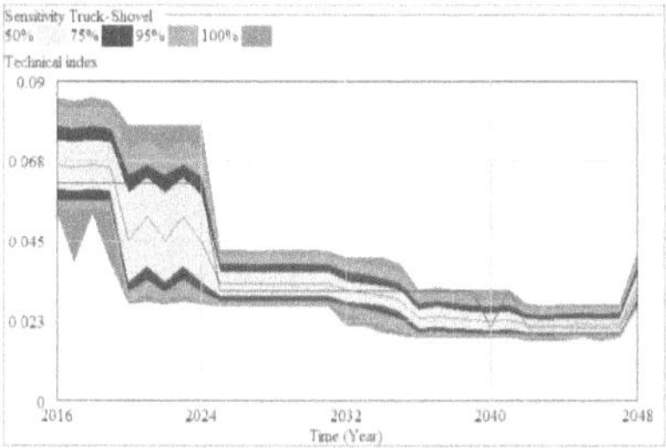

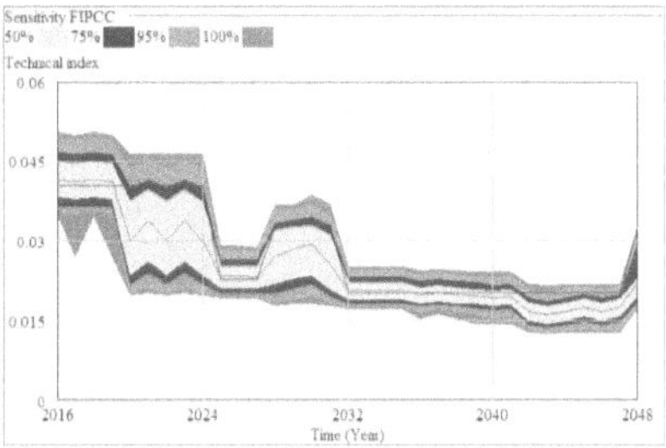

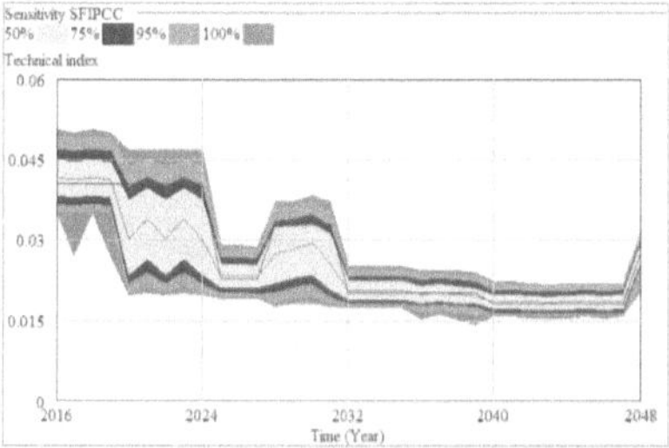

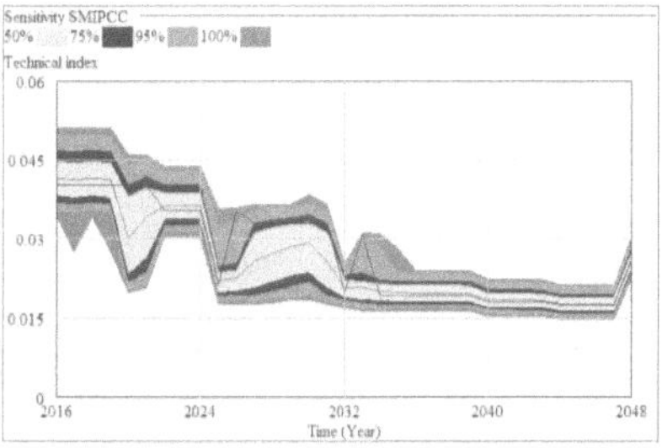

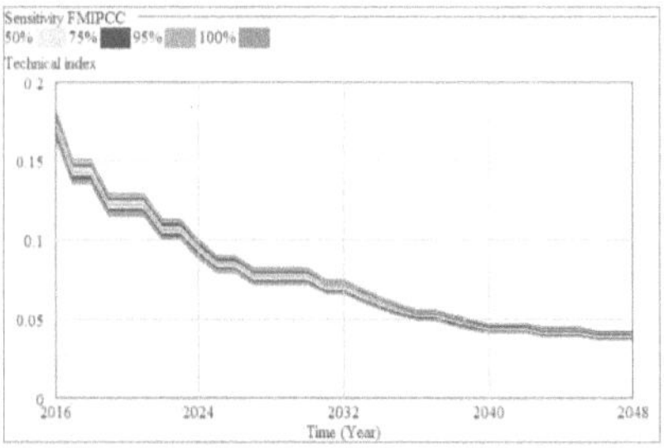

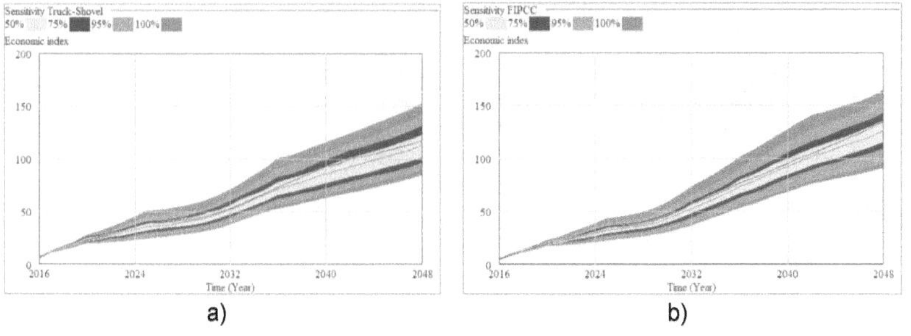

Figure 5-38. Sensitivity analysis for technical index of a) Truck-Shovel b) FIPCC c) SFIPCC d) SMIPCC and e) FMIPCC systems

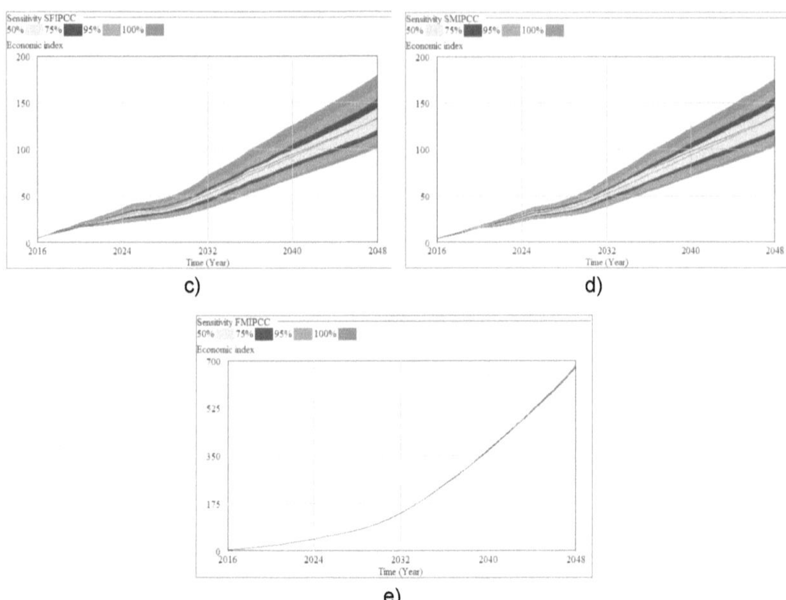

Figure 5-39. Sensitivity analysis for economic index of a) Truck-Shovel b) FIPCC c) SFIPCC d) SMIPCC and e) FMIPCC systems

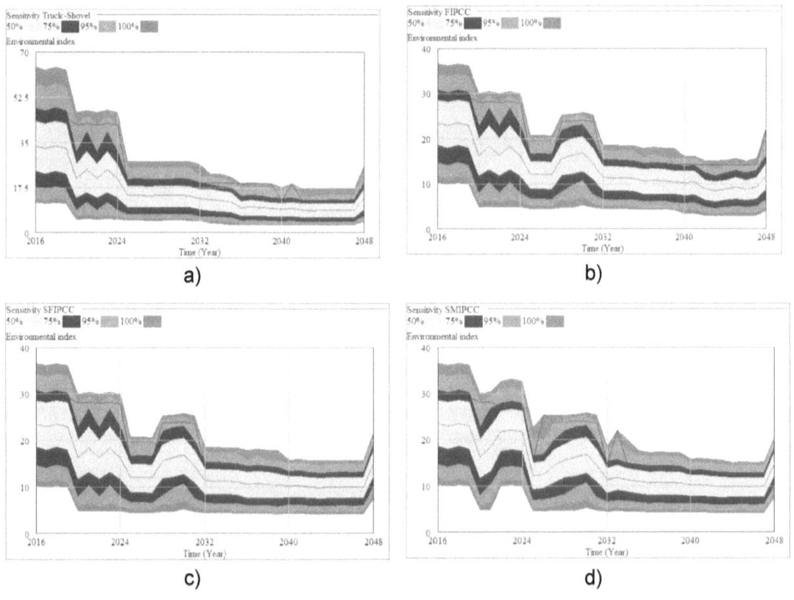

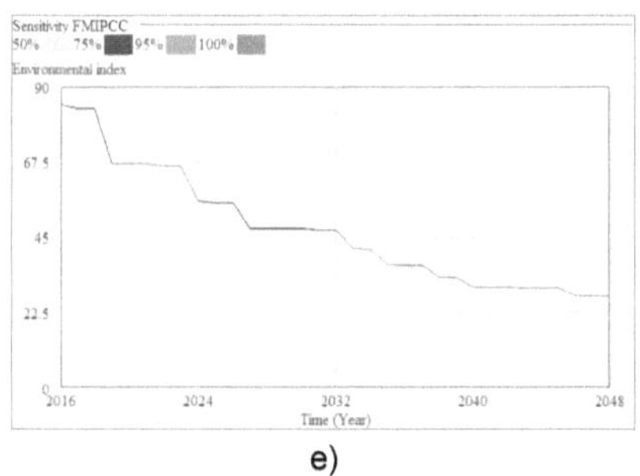

e)

Figure 5-40. Sensitivity analysis for environmental index of a) Truck-Shovel b) FIPCC c) SFIPCC d) SMIPCC and e) FMIPCC systems

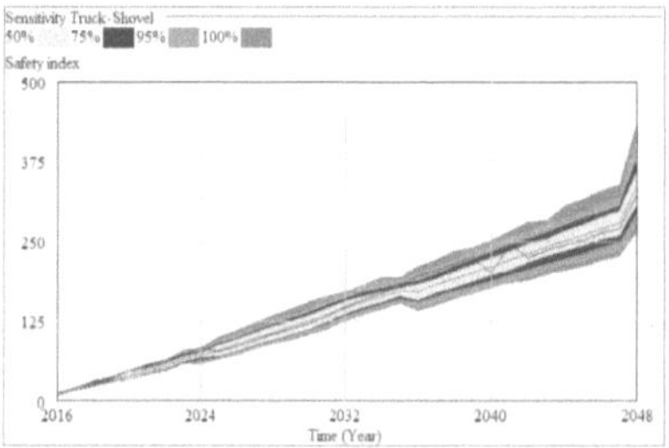

a)

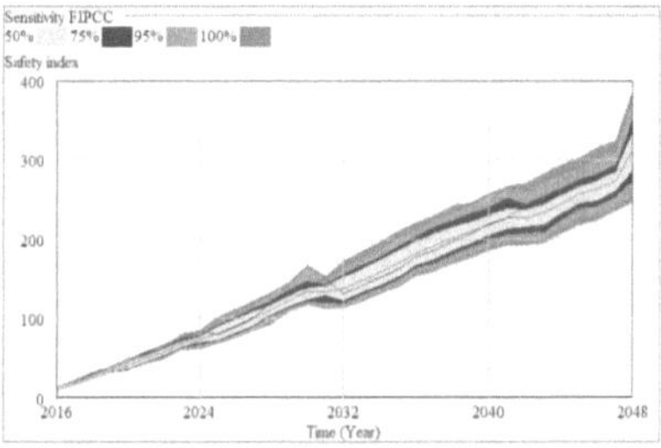

b)

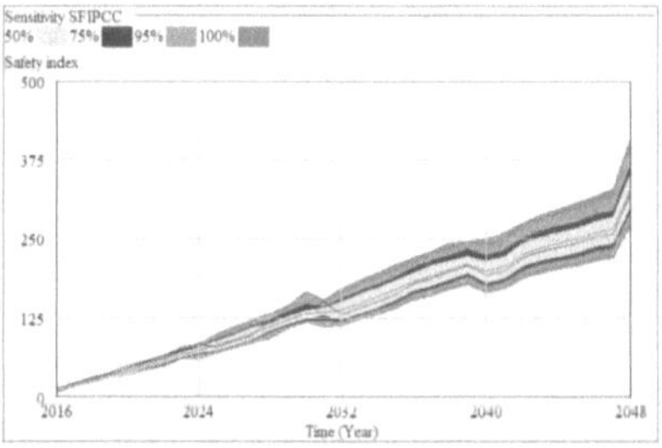

c)

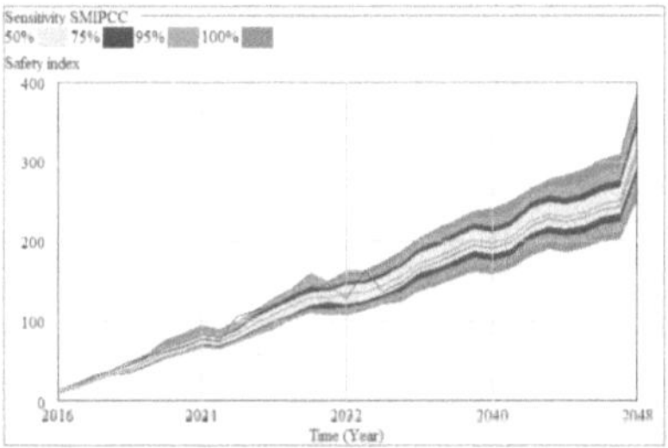

d)

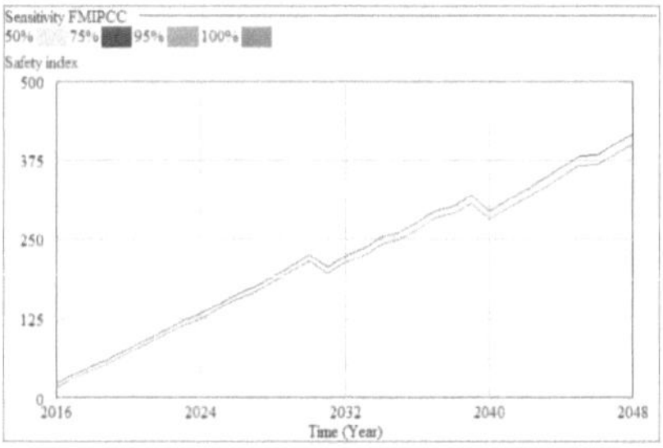

e)

Figure 5-41. Sensitivity analysis for safety index of a) Truck-Shovel b) FIPCC c) SFIPCC d) SMIPCC and e) FMIPCC systems

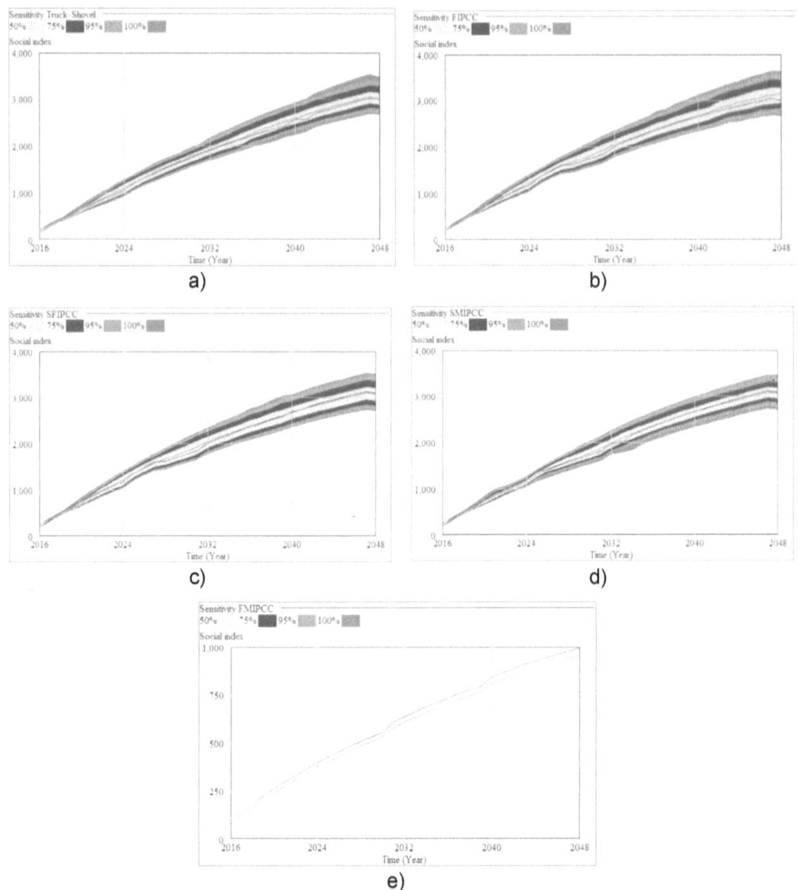

Figure 5-42. Sensitivity analysis for the social index of a) Truck-Shovel b) FIPCC c) SFIPCC d) SMIPCC and e) FMIPCC systems

5.8. Transportation system selection

As described in Sections 0 to 5.6, all the indexes for the different transportation systems and their relative parameters can be determined. However, this would not be solely enough, and it is still needed that the transportation system selection to be performed. Accordingly, Sustainability Index software, which was developed in the Java programming language, was provided. It works based on the AHP process, which finally will result in the transportation system selection. The AHP process was thoroughly explained in Sections 4.2 to 4.5, and in the following, its application and the process in selecting the transportation system will be explained.

5.8.1. Getting output from TEcESaS application for importing in Sustainability software

For selecting the transportation system by Sustainability Index software, the first step, as mentioned in 14.6.1, is to import the output of TEcESaS Indexes software into Sustainability Index software. Accordingly, it is needed to be determined what would be the indexes if any of the transportation systems start in any year of the project. Hence, the following procedure was implemented:

1. A sensitivity analysis on the "start year of the project", indicated by "INITIAL TIME" in the model, was performed in TEcESaS Index software. The distribution function was "vector distribution" with an increment of one, minimum and maximum values of "start year of the ore extraction" and "FINAL TIME", respectively. It will result in a matrix $I_{A_k} = [i_{m,n}]$, which A_k are the alternatives ($k = \{1,..,5\}$) and $i_{m,n}$ is the index of the alternative for the year m with the start year n.

2. The outputs of these sensitivity runs are saved with the name "SA *.tab" in the project, which * can be the name of the transportation system, e.g., "SA FIPCC.tab".

3. The *.tab files are imported into the Sustainability Index software to start selecting the transportation system.

4. In this step, the pairwise comparison of different indexes should be made according to the process explained in Section 4.5.1. Therefore, the comparison values for each index related to the other indexes are imported. For this case study, Table 5-9 as the pairwise comparison is designed.

5. After constructing the pairwise comparison matrix, by pressing the "calculate" button, the software will run the AHP based on the data imported in the previous steps. Finally, the transportation system selection will be determined based on the following algorithm each year:

    5.1. $Best\ Index_n\ (BI_n) = \max\left\{i_{(n,n)A_k}\right\}$

    5.2.   if $BI_n = i_{(n,n)A_1}$, then

    5.3. $BI_{n+1} = \max\left\{i_{(n+1.n)A_1}, i_{(n+1,n+1)A_\alpha}\right\}\ \alpha = \{2,3,4,5\}$

    5.4.   else if $BI_n = i_{(n,n)A_2}$, then

    5.5. $BI_{n+1} = \max\left\{i_{(n+1.n)A_2}, i_{(n+1,n+1)A_\beta}\right\}\ \beta = \{1,3,4,5\}$

    5.6.   else if $BI_n = i_{(n,n)A_3}$, then

    5.7. $BI_{n+1} = \max\left\{i_{(n+1.n)A_3}, i_{(n+1,n+1)A_\gamma}\right\}\ \gamma = \{1,2,4,5\}$

    5.8.   else if $BI_n = i_{(n,n)A_4}$, then

    5.9. $BI_{n+1} = \max\left\{i_{(n+1.n)A_4}, i_{(n+1,n+1)A_\delta}\right\}\ \delta = \{1,2,3,5\}$

    5.10.   else if $BI_n = i_{(n,n)A_5}$, then

    5.11. $BI_{n+1} = \max\left\{i_{(n+1.n)A_5}, i_{(n+1,n+1)A_\varepsilon}\right\}\ \varepsilon = \{1,2,3,4\}$

    5.12.   n = n + 1

    5.13.   if n = m then

    5.14   break

    5.15.   else

    5.16   goto 5.2.

    5.17.   return BI_n

5.8.2. Transportation system selection by single expert (deterministic mode)

In the final step in the Sustainability Index software, in which all files were imported and the pairwise comparison matrix was specified, the software will perform the transportation system selection in the single expert mode by pressing the "Calculate in Single Expert Mode" button. It is shown in Figure 5-43.

Table 5-9. Pairwise comparison matrix for different indexes

	Technical index	Economic index	Environmental index	Safety index	Social index
Technical index	1	0.2	1	1	0.5
Economic index	5	1	9	8	6
Environmental index	1	0.111	1	1	2
Safety index	1	0.125	1	1	1
Social index	2	0.167	0.5	1	1

As it can be seen in Figure 5-43a, the highest sustainability index from 2016 and 2017 belongs to the Truck-Shovel system. Accordingly, in these years, this system should be applied. Nevertheless, from 2018 to 2048, the FMIPCC system shows the best sustainability index (Figure 5-43a, b, and c), which nominates this system as the selected transportation system for the rest of the project. Figure 5-43d depicts the graphical view of the transportation system selection. The table related to these figures is presented in Appendix IV.

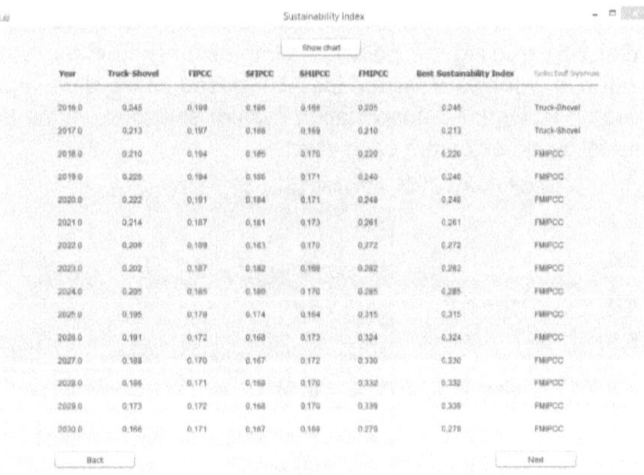

a)

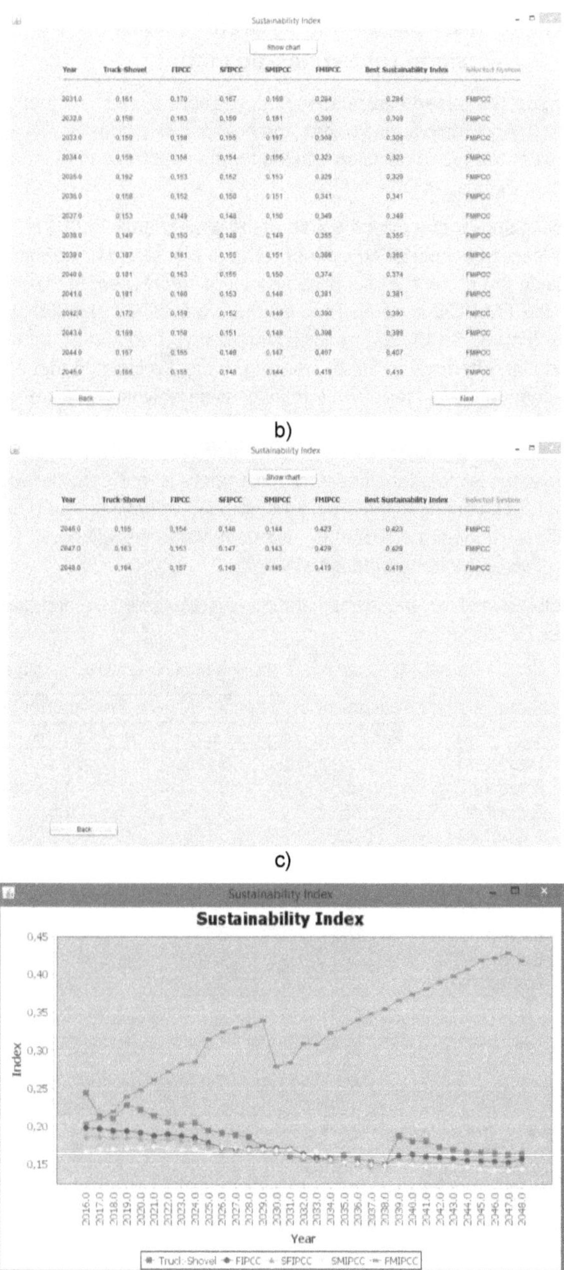

Show chart

Year	Truck-Shovel	FIPCC	SFIPCC	SMIPCC	FMIPCC	Best Sustainability Index	Selected System
2031.0	0.161	0.170	0.167	0.169	0.284	0.284	FMIPCC
2032.0	0.158	0.163	0.159	0.181	0.309	0.309	FMIPCC
2033.0	0.159	0.198	0.155	0.197	0.308	0.308	FMIPCC
2034.0	0.159	0.156	0.154	0.156	0.323	0.323	FMIPCC
2035.0	0.192	0.153	0.152	0.153	0.329	0.329	FMIPCC
2036.0	0.158	0.152	0.150	0.151	0.341	0.341	FMIPCC
2037.0	0.153	0.149	0.148	0.150	0.349	0.349	FMIPCC
2038.0	0.149	0.150	0.148	0.149	0.355	0.355	FMIPCC
2039.0	0.187	0.161	0.155	0.151	0.366	0.366	FMIPCC
2040.0	0.181	0.163	0.155	0.150	0.374	0.374	FMIPCC
2041.0	0.181	0.160	0.153	0.148	0.381	0.381	FMIPCC
2042.0	0.172	0.159	0.152	0.149	0.390	0.390	FMIPCC
2043.0	0.169	0.158	0.151	0.148	0.398	0.398	FMIPCC
2044.0	0.167	0.155	0.148	0.147	0.407	0.407	FMIPCC
2045.0	0.166	0.155	0.148	0.144	0.419	0.419	FMIPCC

Back   Next

b)

Show chart

Year	Truck-Shovel	FIPCC	SFIPCC	SMIPCC	FMIPCC	Best Sustainability Index	Selected System
2046.0	0.155	0.154	0.148	0.144	0.423	0.423	FMIPCC
2047.0	0.163	0.151	0.147	0.143	0.429	0.429	FMIPCC
2048.0	0.164	0.157	0.149	0.145	0.419	0.419	FMIPCC

Back

c)

d)

Figure 5-43. The transportation system selection through the mine's life for a) 2016-2030 b) 2031-2045 c) 2046-2048 and d) graphical view

5.8.3. Transportation system selection by group decision making (deterministic mode)

In this mode, the experts' evaluation matrix is filled by ten hypothetical experts, shown in Table 5-10. According to this matrix, the final pairwise comparison will be as Table 5-11. The result of the group decision making for the best transportation system in the project is shown in Figure 5-44.

Figure 5-44 shows that from the start of the project until 2020, the Truck-Shovel system is the selected transportation system with the highest sustainability index. However, by progressing the project, its sustainability index decreased. From 2021 to the end of the project, the FMIPCC is determined as the project's transpiration system. As it can be seen in Figure 5-44d, its trend is increasing during the project while the other transportation alternatives are decreasing. It can be claimed that although the FMIPCC is not recommended in the short term, it considerably shows a better sustainability index in the long term. However, this is naturally dependent on the inputs and the comparison matrix defined in this case. In the deterministic mode, all the inputs are considered fixed throughout the simulation, and any uncertainty is not considered. However, the real situation is not working in this way, and there would definitely be a degree of uncertainty. Accordingly, running the model in uncertain situations will provide a better envision of the selection of the transportation system.

The table related to the transportation system selection indexes can be found in Appendix IV.

Table 5-10. Experts' evaluation matrix in the group decision making

Expert ID	Technical / Economic	Economic / Environmental	Environmental / Safety	Safety / Social
Expert #1	0.2	9	0.333	0.5
Expert #2	0.167	8	0.2	1
Expert #3	0.25	8	0.5	0.333
Expert #4	0.2	6	1	0.167
Expert #5	1	7	0.333	0.5
Expert #6	1	1	1	1
Expert #7	0.2	6	0.25	0.333
Expert #8	1	6	0.333	0.25
Expert #9	0.25	9	1	0.5
Expert #10	0.333	9	0.2	0.25

Table 5-11. Comparison matrix for the group decision making (10 experts)

	Technical index	Economic index	Environmental index	Safety index	Social index
Technical index	1	0.333	2	1	0.333
Economic index	3	1	6	3	1
Environmental index	0.5	0.167	1	0.5	0.167
Safety index	1	0.333	2	1	0.333
Social index	3	1	6	3	1

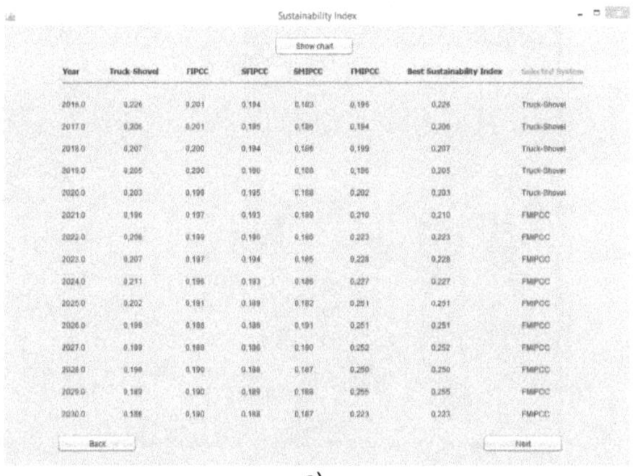

Sustainability Index

Show chart

Year	Truck-Shovel	FIPCC	SFIPCC	SMIPCC	FMIPCC	Best Sustainability Index	Selected System
2016.0	0.226	0.201	0.194	0.183	0.196	0.226	Truck-Shovel
2017.0	0.206	0.201	0.195	0.186	0.194	0.206	Truck-Shovel
2018.0	0.207	0.200	0.194	0.186	0.199	0.207	Truck-Shovel
2019.0	0.205	0.200	0.196	0.188	0.186	0.205	Truck-Shovel
2020.0	0.203	0.199	0.195	0.188	0.202	0.203	Truck-Shovel
2021.0	0.196	0.197	0.193	0.189	0.210	0.210	FMIPCC
2022.0	0.206	0.199	0.196	0.186	0.223	0.223	FMIPCC
2023.0	0.207	0.197	0.194	0.185	0.228	0.228	FMIPCC
2024.0	0.211	0.196	0.193	0.186	0.227	0.227	FMIPCC
2025.0	0.202	0.191	0.189	0.182	0.251	0.251	FMIPCC
2026.0	0.199	0.188	0.186	0.191	0.251	0.251	FMIPCC
2027.0	0.199	0.188	0.186	0.190	0.252	0.252	FMIPCC
2028.0	0.198	0.190	0.188	0.187	0.250	0.250	FMIPCC
2029.0	0.189	0.190	0.189	0.186	0.255	0.255	FMIPCC
2030.0	0.188	0.190	0.188	0.187	0.223	0.223	FMIPCC

Back   Next

a)

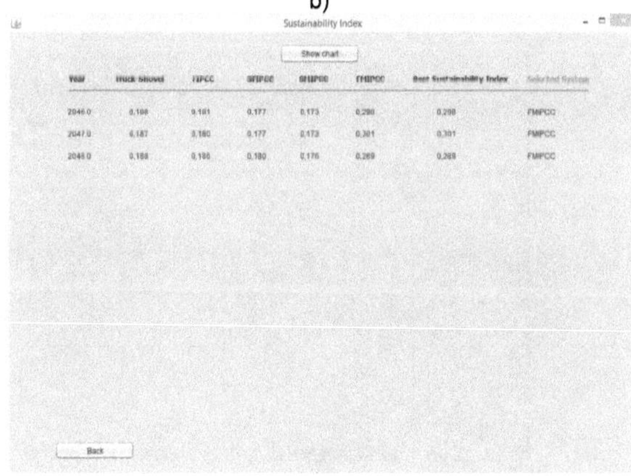

Sustainability Index

Show chart

Year	Truck-Shovel	FIPCC	SFIPCC	SMIPCC	FMIPCC	Best Sustainability Index	Selected System
2031.0	0.183	0.190	0.188	0.188	0.223	0.223	FMIPCC
2032.0	0.178	0.186	0.183	0.182	0.245	0.245	FMIPCC
2033.0	0.181	0.181	0.180	0.188	0.239	0.239	FMIPCC
2034.0	0.182	0.182	0.180	0.179	0.248	0.248	FMIPCC
2035.0	0.184	0.180	0.179	0.177	0.250	0.250	FMIPCC
2036.0	0.183	0.180	0.179	0.178	0.256	0.256	FMIPCC
2037.0	0.180	0.178	0.177	0.177	0.259	0.259	FMIPCC
2038.0	0.178	0.178	0.178	0.176	0.261	0.261	FMIPCC
2039.0	0.202	0.184	0.181	0.177	0.269	0.269	FMIPCC
2040.0	0.199	0.186	0.180	0.175	0.273	0.273	FMIPCC
2041.0	0.201	0.184	0.179	0.175	0.276	0.276	FMIPCC
2042.0	0.194	0.184	0.179	0.176	0.281	0.281	FMIPCC
2043.0	0.182	0.183	0.179	0.176	0.283	0.283	FMIPCC
2044.0	0.187	0.183	0.179	0.175	0.289	0.289	FMIPCC
2045.0	0.188	0.182	0.177	0.173	0.296	0.296	FMIPCC

Back   Next

b)

Sustainability Index

Show chart

Year	Truck-Shovel	FIPCC	SFIPCC	SMIPCC	FMIPCC	Best Sustainability Index	Selected System
2046.0	0.188	0.181	0.177	0.173	0.298	0.298	FMIPCC
2047.0	0.187	0.180	0.177	0.173	0.301	0.301	FMIPCC
2048.0	0.188	0.186	0.180	0.176	0.289	0.289	FMIPCC

Back

c)

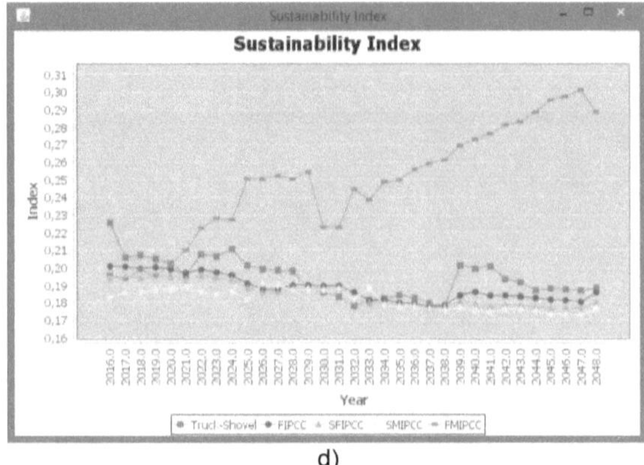

d)

Figure 5-44. The result of the selected transportation system in group decision-making mode

5.8.4. Transportation system selection by group decision making (stochastic mode)

For selecting the transportation system in the stochastic mode, a Monte Carlo simulation on the constants described in Table 5-8 and based on the specified distribution was performed. In contrast with the deterministic mode, in which these constants were fixed throughout the simulation in TEcESaS Indexes software, in the stochastic mode, the results were calculated after 10000 simulations based on the defined distributions for the inputs (Table 5-8). The result shows that the FMIPCC, which has the highest probability (79%-92%) among the other transportation alternatives, should be selected as the transportation system throughout the mine's life (Figure 5-45). At the second rank, the Truck-Shovel system was placed with a probability of 6%-21%. Moreover, in third place stood the FIPCC system with a probability of 0%-6%. These numbers are presented as a form of a table in Appendix IV.

a)

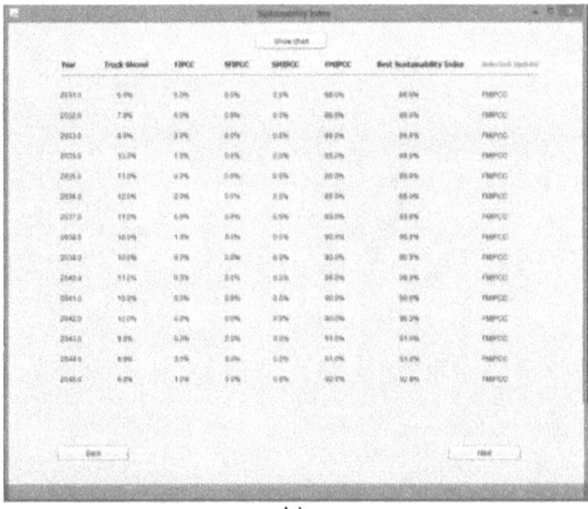

b)

c)

Figure 5-45. Transportation system selection in the stochastic mode

CHAPTER 6: Summary and recommendations

6.1. Summary

Nowadays, and by progressing industries, which are daily demanding for more raw materials as their fundamental need to make final products, the mining industry as their provider undoubtedly requires adjusting itself with the burgeoning demands. This will not be fulfilled unless introducing new mining methods to catch up with fast-growing changes.

The transportation system in the mining projects, especially in open-pit mines, which is responsible for transferring daily tens or hundreds of thousands of materials to the desired destinations, plays a key role in all the mining process. The truck-shovel system, as the most recognized transportation system in open-pit mines, has been using for many decades. Despite its high flexibility and relatively low capital cost, it is counted as a non-environmentally friendly and unsafe system due to burning fuels and accidents, respectively. IPCC systems, which benefit from conveyor belts for transferring material, are the other types of transportation systems. Unlike some cons such as high capital cost and low flexibility, their pros are including lower capital cost, more environmentally friendly, and higher safety.

Although many studies have been conducted related to these transportation systems, they are mainly discussing an individual part of the problem. Some of them are technical perspective (e.g., final pit design, flexibility, availability, and utilization, etc.), economic issues (e.g., capital and operating costs), and environmental aspects (e.g., fuel and electricity consumption). As it is evident from the literature, each of them merely notices one side of the transportation system and does not contribute to another part. For instance, the safety aspect's contribution is few while there is no literature about social viewpoint. Additionally, they are mostly discussing the present situations in a statistical approach and does not offer any dynamic solution. All these facts will result in a situation that mine managers will not be able to make a proper decision about selecting the superior transportation system.

According to the above-mentioned challenges regarding the transportation system, it is highly in demand to propose a way that not only provides the possibility to compare all transportation system alternatives on all the aspects but also gives an insight to the managers to help for making right decisions. Accordingly, the following objectives were defined for this research:

- Defining a dynamic model of the transportation system in open-pit mines by considering the Truck-Shovel and IPCC systems.
- Describing the technical, economic, environmental, safety and social indexes for these transportation systems in the dynamic model.
- Comparing these systems regarding their outcomes in the indexes mentioned above.
- Software development, as a result of building dynamics model, for presenting to the public.
- Software development based on AHP, using the output of Vensim software as input, performs the transportation system selection through the mine's life.

To fulfill these objectives, the following steps were taken in this study:

As the first step, different technical, economic, environmental, safety, and social issues were investigated in a system dynamics approach. Accordingly, each of them was modeled in system dynamics through Vensim software. The modeling process was included finding as much as parameters in any issue, the relations, equations, feedback, feedback loops, etc., which finally one unique index could be defined for any issues. These indexes are technical index (TI), economic index (EcI), environmental index (EI), safety index (SaI) and social index (SI). As the final action in this step, an application was provided using Venapp. It was named "TEcESaS Indexes", which any user can efficiently utilize to get the output of different parameters and indexes in Truck-Shovel and IPCC systems.

As the second step, the transportation system selection was performed based on the output from the first step through the AHP. This process was included defining an overall index for any transportation system, which embraces all the technical, economic, environmental, safety, and social indexes. This index, which was named "Sustainability Index", was written in Java programming language, and its relevant software was developed. A final decision could be made through this software that which of the transportation system options can be chosen regarding the highest sustainability index along with the mine's life.

Finally, and as the case study, a hypothetical copper mine was considered, and the model was performed in different states of single expert (deterministic and stochastic) and group decision making (stochastic mode). The results showed that in the deterministic simulation, the best transportation system in the single expert mode was the Truck-Shovel system for 2016 and 2017 and the FMIPCC system from 2018 to the end of the project. Additionally, in the group decision deterministic mode, Truck-Shovel will be the best transportation system from 2016 to 2020. However, after this year, and to the end of the project, the FMIPCC will be the selected system. Unlike the deterministic mode, the FMIPCC with the highest probability should be selected as the mine's transportation system in the group decision stochastic mode.

6.2. Recommendations

This research could fill the gaps from previous studies regarding the transportation systems to a great extent. The most important of them were considering different technical, economic, environmental, safety, and social factors as an integrated system through system dynamics modeling. It provided the ability to choose the best transportation system through the mine life by considering all these indexes. Furthermore, it made it possible to compare all these alternatives in various parameters and indexes above. Nevertheless, there are still opportunities for further works in this regard that the most significant of them are mentioned as follows:

➢ This study merely focused on the haulage system in open-pit mines. Since there are other significant operations in mining projects such as drilling and blasting, it would be helpful if they are also modeled in the frame of system dynamics modeling like this book process. The connection between these two operations can provide a broader viewpoint for the mine project and its status along with mine's life.

➢ Trying other methods rather than system dynamics and AHP, e.g., fuzzy AHP etc., for modeling and selecting the best transportation system and its comparison with these methods will be another interesting point for further studies. However, the indexes of this study should be taken into account and, additionally, being able to provide a final decision of choosing the transportation system.

➢ This study presented the problem in the condition that just one IPCC system is operating in the mine. This assumption can be extended into more comprehensive options, such as considering more than one IPCC system and even combining different types of IPCCs. This case will help mega and very deep open-pit mines, which probably many crushing stations should be installed.

➢ Since the mine production scheduling would be different for Truck-Shovel and IPCC systems, as mentioned in the literature review, it would be interesting if the application and software provided in this study could be coupled with ones developed for production scheduling of IPCCs. It will positively strengthen the making decisions of selecting the proper transportation system.

References

[1] "Mtlexs.com," Mtlexs.com, [Online]. Available: http://www.mtlexs.com/demand-supply/refined-copper-production-and-usage/4.

[2] "Ende Gelände 2017," [Online]. Available: https://www.ende-gelaende.org/en/.

[3] Oracle Mining Corp., "Oracle Mining Corp.," [Online]. Available: http://www.oracleminingcorp.com/copper/.

[4] "Macrotrends," [Online]. Available: http://www.macrotrends.net/1476/copper-prices-historical-chart-data.

[5] R. M. Hays, "Mine planning considerations for in-pit crushing and conveying systems," *SME Mini Symposium,* pp. 33-41, 1983.

[6] E. M. Frizzell, "Mobile In-Pit Crushing - Product of Evolutionary Change," *Trans. Am. Inst. Mining, Metall. Pet. Eng. Soc.,* pp. 578-580, 1985.

[7] R. Ritter, "Contribution to the capacity determination of semi-mobile in-pit crushing and conveying systems," Quality Content of Saxony (qucosa), TU Bergakademie Freiberg, 2016.

[8] F. Koehler, "Inpit crushing system the future mining option," in *Twelfth International Symposium on Mine Planning and Equipment Selection (MPES 2003),* Kalgoorlie, Western Australia , 2003.

[9] D. Turnbull and A. Cooper, "In-pit crushing and conveying (IPCC)–a tried and tested alternative to trucks," *AusIMM Bulletin,* pp. 60-64, 2010.

[10] D. Tutton and W. Streck, "The application of mobile in-pit crushing and conveying in large, hard rock open pit mines," in *Mining Magazine Congress,* Canada, 2009.

[11] E. Zimmermann and W. Kruse, "Mobile crushing and conveying in quarries - a chance for better and cheaper production!," in *8th International Symposium Continuous Surface Mining,* Aachen, 2006.

[12] M. Osanloo and M. Paricheh, "In-pit crushing and conveying technology in open pit mining operations: a literature review and research agenda," *International Journal of Mining, Reclamation and Environment,* pp. 1-28, 2019.

[13] T. Atchison and D. Morrison, "In-pit crushing and conveying bench operations," in *Iron Ore Conference,* Pert, WA, Australia, 2011.

[14] M. Johnson, "Impact of in-pit crushing and conveying on pit shell optimization," Deswik Mining Consultants, Australia, 2014.

[15] E. Hay, M. Nehring, P. Knights and M. S. Kizil, "Ultimate pit limit determination for semi mobile in-pit crushing and conveying system: a case study," *International Journal of Mining, Reclamation and Environment,* 2019.

[16] E. Hay, Ultimate Pit Limit Determination for Fully Mobile In-Pit Crushing and Conveying Systems, Queensland: The University of Queensland, 2018.

[17] M. Dean, P. Knights, M. S. Kizil and M. Nehring, "Selection and planning of fully mobile in-pit crusher and conveyor systems for deep open pit metalliferous applications," in *Third international future mining conference*, NSW, Sydney, 2015.

[18] N. Nehring, P. Knights, M. Kizil and E. Hay, "A comparison of strategic mine planning approaches for in-pit crushing and conveying, and truck/shovel systems," *International Journal of Mining Science and Technology,* vol. 28, pp. 205-214, 2018.

[19] J. Sturgul, "How to determine the optimum location of in-pit movable crusher," *International journal of mining and geological engineering,* pp. 143-148, 1987.

[20] Y. Changzhi, "In-pit crushing and conveying system in Dexin pit copper haulage optimization for ore transport," in *Fifth Large Open Pit Mining Conference,* Kalgoorlie, 2003.

[21] G. Konak, "Selection of the optimum in-pit crusher location for an aggregate producer," *The Journal of the Southern African,* pp. 161-166, 2007.

[22] M. Rahmanpour, M. Osanloo, N. Adibee and M. AkbarpourShirazi, "An approach to locate an in-pit crusher in open pit mines," *International Journal of Engineering,* pp. 1475-1484, 22 May 2014.

[23] M. Paricheh, M. Osanloo and M. Rahmanpour, "A heuristic approach for in-pit crusher and conveyor system's time and location problem in large open-pit mining," *International Journal of Mining, Reclamation and Environment,* 8 October 2016.

[24] M. Paricheh and M. Osanloo, "Determination of the optimum in-pit crusher location in open-pit mining under production and operating cost uncertainties," Istanbul, Turkey, 2016.

[25] H. Abbaspour, C. Drebenstedt, M. Paricheh and R. Ritter, "Optimum location and relocation plan of semi-mobile in-pit crushing and conveying systems in open-pit mines by transportation problem," *International Journal of Mining, Reclamation and Environment,* pp. 1-21, 2018.

[26] C. A. J. Builes, A Mixed-Integer Programming Model for an In-Pit Crusher Conveyor Location Problem, Montreal, Canada: Univerite de Montreal, Ecole Polytechnic de Montreal, 2017.

[27] P. Morriss, "Key production drivers in in-pit crushing and conveying (IPCC) studies," *The Southern African Institute of Mining and Metallurgy,* pp. 23-34, 2008.

[28] B. McCarthy, "Evaluating the lack of flexibility against the benefits of in-pit crushing and conveying," in *Optimizing Mine Operations Conference*, Toronto, 2013.

[29] I. Dzakpata, P. Knights, M. S. Kizil, M. Nehring and S. M. Aminossadati, "Truck and Shovel Versus In-Pit Conveyor Systems: a Comparison Of The Valuable Operating Time," in *Coal Operators' Conference*, University of Wollongong, 2016.

[30] J. A. Dos Santos and Z. Stanisic, "In-pit crushing and high angle conveying in Yugoslavian copper mine," *Internation Journal of Surface Mining 1,* pp. 97-104, 1987.

[31] M. Paricheh and M. Osanloo, "How to Exit Conveyor from an Open-Pit Mine: A Theoretical Approach," in *Proceedings of the 27th International Symposium on Mine Planning and Equipment Selection - MPES 2018*, Cham, Switzerland, Springer, 2019, pp. 319-334.

[32] B. A. Kammerer, "In-pit crushing and conveying system at Bingham Canyon Mine," *International Journal of Surface Mining 2,* pp. 143-147, 1988.

[33] J. K. Radlowski, "In-pit crushing and conveying as an alternative to an all truck system in open pit mines," The University of British Columbia, Cracow, Poland, 1988.

[34] T. W. Martin, "Large rock conveying systems and their application in open pit mines," in *Mini Sysmposium: Open pit haulage systems in the 80's, SME AIME anuual meeting*, Chicago, 1981.

[35] R. W. Utley, "Component analysis for movable in-pit crushers," in *AIME 44th annual mining symposium, 56th annual meeting of the Minnesota section*, Dulluth, Minn, 1983.

[36] N. Terezopoulos, "Continuous haulage and in-pit crushing in surface mining," *Mining Science and Technology,* pp. 253-263, 1988.

[37] H. Lieberwirth, "Economic advantages of belt conveying in open-pit mining," in *Mining Latin America*, Dordrecht, Springer, 1994, pp. 279-295.

[38] C. M. de Almeida, T. de Castro Neves, C. Arroyo and P. Campos, "Proceedings of the 27th International Symposium on Mine Planning and Equipment Selection - MPES 2018," in *Truck-and-Loader Versus Conveyor Belt System: An Environmental and Economic Comparison*, Cham, Switzerland, Springer, 2019, pp. 307-318.

[39] R. A. Nunes, H. D. Junior, G. de Tomi, C. B. Infante and B. Allan, "A decision-making method to assess the benefits of a semi-mobile in-pit crushing and conveying alternative during the early stages of a mining project," *REM: International Engineering Journal,* vol. 72, no. 2, pp. 285-291, 2019.

[40] T. Norgate and N. Haque, "The greenhouse gas impact of IPCC and ore-sorting technologies," *Minerals Engineering,* pp. 13-21, 2013.

[41] V. Raaz and U. Mentges, "Comparison of energy efficiency and CO2 emissions for trucks haulage vs in-pit crushing and conveying of materials: calculation methods and case studies," in *SME Annual Meeting*, Denver, 2011.

[42] K. Awuah-Offei, D. Checkel and H. Askari-Nasab, "Evaluation of belt conveyor and truck haulage systems in an open pit mine using life cycle assessment," *CIM Bulletin,* vol. 4, 2009.

[43] V. Kecojevic, D. Komljenovic, W. Groves and M. Radomsky, "An analysis of equipment-related fatal accidents in U.S. mining operations: 1995–2005," *Safety Science,* vol. 45, pp. 864-874, 2007.

[44] J. Hill, "An assessment of the effectiveness of safety interventions in the field of bulk material handling," in *International Materials Handling Conference*, Pretoria, 2011.

[45] R. Ritter, A. Herzog and C. Drebenstedt, "Automated Dozer Concept Aims to Cut IPCC Downtime," *E&MJ,* pp. 52-55, 2014.

[46] J. K. Radlowski, "In-Pit Crushing and Conveying as an Alternative to an All Truck System in Open Pit Mines," The University of British Columbia, Cracow, Poland, 1988.

[47] P. Darling and R. W. Utley, "In-Pit Crushing," in *SME Mining Engineering Handbook*, USA, Society for Mining, Metallurgy, and Exploration, Inc., 2011, pp. 941-956.

[48] D. Turnbull, "How do you make an IPCC system function," Sandvik Mining Systems, 2013.

[49] TAKRAF, "Mining Technology," [Online]. Available: https://www.mining-technology.com/contractors/highwall/takraf/attachment/takraf4/. [Accessed 05 02 2020].

[50] TAKRAF, "TAKRAF TENOVA," [Online]. Available: https://www.takraf.tenova.com/product/mobile-conveyor-bridges/. [Accessed 05 02 2020].

[51] Metso, "metso," [Online]. Available: https://www.metso.com/products/conveyors/mobile-conveyors/. [Accessed 05 02 2020].

[52] thyssenkrupp, "thyssenkrupp," [Online]. Available: https://www.thyssenkrupp-industrial-solutions.com/en/products-and-services/mining-systems/belt-conveyors. [Accessed 05 02 2020].

[53] T. Schools, "Condition monitoring of critical mining conveyors," *Engineering and Mining Journal,* 03 2015.

[54] "FAM: Förderanlagen Magdeburg," [Online]. Available: http://www.fam.de/english/News/newsarchive/news.52.html.

[55] E. Pruyt, Small System Dynamics Models for Big Issues, Netherlands: TU Delft Library, Delft, The Netherlands, 2013.

[56] B. Kanti Bala, F. Mohamed Arshad and K. Mohd Noh, System Dynamics Modelling and Simulation, Singapore: Springer, 2017.

[57] M. J. Radzicki and R. A. Taylor, Introduction to System Dynamics, A Systems Approach to Understanding Complex Policy Issues, Sustainable Solutions, Inc., 1997.

[58] Y. Barlas, System Dynamics, Encyclopedia of Life Support System, 2009.

[59] S. Albin, "Building a System Dynamics Model Part 1: Conceptualization," Massachusetts Institute of Technology, Cambridge, 1997.

[60] T. R. Binder, A. Vox, S. Belyazid, H. Haraldsson and M. Svensson, "Developing system dynamics models from casual loop diagrams," in *Semantic Scholar*, 2004.

[61] S. Elevli and B. Elevli, "Performance measurement of mining equipments by utilizing OEE," *Acta Montanistica Slovaca,* vol. 15, pp. 95-101, 2010.

[62] H. Rohani and A. Kamali Roosta, "Calculating Total System Availability," Information Services Organization KLM-Air France, Amsterdam, 2014.

[63] C. Shu-zhao, C. Qing-xiang, Z. Wei and Z. Lei, "Study on new pattern of semi-continuous mining system used in surface mines," *Procedia Earth and Planetary Science,* vol. 1, pp. 243-249, 2009.

[64] A. Soofastaei, S. M. Aminossadati, M. S. Kizil and P. Knights, "Reducing fuel consumption of haul trucks in surface mines using artificial intelligence models," in *Coal Operators' Conference*, Wollongong, Australia, 2016.

[65] O. Leiva, "Haulage Profile Advances in Vulcan 9," MAPTEK, Sydney, Australia, 2013.

[66] I. Assakkaf, "Machine Power," in *Construction Equipment and Mehods*, Maryland, University of Maryland, 2003.

[67] CEMA, "Belt Conveyors for Bulk Materials," in *Belt Tension, Power, and Drive Engineering*, 6 ed., Conveyor Equipment Manufacturers Association, 2007, pp. 85-196.

[68] USEPA, Compilation of air pollutant emission factors, North Carolina: United States Environmental Protection Agency, 1995.

[69] SPCC, Air pollution from coal mining and related developments, Sydney: State Pollution Control Commission, 1983.

[70] NERDC, Air Pollution from Surface Coal Mining: Volume 2 Emission Factors and Model Refinement, National Energy Research and Demonstration Council, 1988.

[71] EIA, "U.S. Energy Information Administration," 10 May 2017. [Online]. Available: https://www.eia.gov/tools/faqs/faq.php?id=107&t=3.

[72] EIA, "Annual electric generator report," U.S. Energy Information Administration, 2015.

[73] EIA, " Monthly energy review," U.S. Energy Information Administration , 2017.

[74] V. Quaschning, Regenerative energiesysteme, München: Carl Hanser, 2015.

[75] Eastern Research Group, Inc., "Uncontrolled emission factor listing for criteria air pollution," in *Emission Inventory Improvement Program (EIIP)*, United States Environental Protection Agency, 2001.

[76] EPA, "Average carbon dioxide emission resulting from gasoline and diesel fuel," Environmental Protection Agency, Washington, 2005.

[77] EPA, "Compilation of air pollutant emission factors," Environmental Protection Agency, Ann Arbor, 1985.

[78] M. Vallius, Characteristics and sources of fine particulate matter in urban air, Kuopio: The National Public Health Institute , 2005.

[79] EPA, "Crushed Stone Processing and Pulverized Mineral Processing," in *Compilation of air pollutant emission factors*, North Carolina, U.S. Environmental Protection Agency, 1995.

[80] EPA, "Miscellaneous Sources," in *Compilation of air pollutant emission factors*, North Carolina, U.S. Environmental Protection Agency, 1995.

[81] C. Cowherd, G. E. Muleski and J. S. Kinsey, "Control of Open Fugitive Dust Sources," U. S. Environmental Protection Agency, North Carolina, 1988.

[82] H. Steven, "Investigations on Noise Emission of Motor Vehicles in Road Traffic," German Environmental Agency (UBA), Wuerselen, Germany, 2005.

[83] E. M. Salomons, Computational Atmospheric Acoustics, Dordrecht, Netherlands: Kluwer Academic Publishers, 2001.

[84] A. Gladysiewicz,, "Noise emissions of belt conveyors," in *Coal International*, 2011.

[85] S. C. Brown, "Conveyor noise specification and control," in *Proceeding of ACOUSTICS*, Gold Coast, Australia, 2004.

[86] C. Chew, "Lost-time work accidents in an industry," *The New Zealand Medical Journal,* vol. 97, pp. 564-567, 1984.

[87] J. S. Oh, C. Oh, S. G. Ritchie and M. Chang, "Real-time estimation of accident likelihood for safety enhancement," *Journal of transportation engineering,* vol. 131, pp. 358-363, 2005.

[88] R. Elvik, "The power model of the relationship between speed and road safety," Norwegian Centre for Transportation Research, Oslo, 2009.

[89] S. Kral, "Improved training reduces worker injuries," *Mining Engineering,* vol. 54, pp. 23-26, 2002.

[90] L. Mancini and S. Sala, "Social impact assessment in the mining sector: Review and comparison of indicators frameworks," *Resources Policy,* vol. 57, pp. 98-111, 2018.

[91] A. Azapagic, "Developing a framework for sustainable development indicators for the mining and minerals industry," *Journal of Cleaner Production,* vol. 12, pp. 639-662, 2004.

[92] G. Nilsson, "The effects of speed limits on traffic accidents in Sweden," in *Proceedings of the International Symposium on the Effects of Speed Limits on Traffic Accidents and Transport Energy use,* Dublin, 1981.

[93] D. Richards and R. Cuerden, "The relationship between speed and car driver injury severity," Department for Transport, Transport Research Laboratory, London, 2009.

[94] R. Elvik, "A re-parameterisation of the Power Model of the relationship between the speed of traffic and the number of accidents and accident victims," *Accident Analysis & Prevention,* vol. 50, pp. 854-860, 2012.

[95] C. G. Drury, W. L. Porter and P. G. Dempsey, "Patterns in mining haul truck accidents," in *56th Annual Meeting of the Human Factors and Ergonomics Society* , Boston, 2012.

[96] S. R. Dindarloo, J. P. Pollard and E. Siami-Irdemoos, "Off-road truck-related accidents in U.S. mines," *Journal of Safety Research,* vol. 58, pp. 79-87, 2016.

[97] M. Abdel-aty, A. Pande, C. Lee, V. Gayah and C. D. Santos, "Crash risk assessment using intelligent transportation systems data and real-time intervention strategies to improve safety on freeways," *Journal of Intelligent Transportation Systems,* vol. 11, pp. 107-120, 2007.

[98] S. P. Hoogendoorn, "Traffic flow theory and simulation," Delft University of Technology, Delf, 2007.

[99] A. Theofilatos and G. Yannis, "A review of the effect of traffic and weather characteristics on road safety," *Accident Analysis & Prevention,* vol. 72, pp. 244-256, 2014.

[100] K. Duivenvoorden, "The relationship between road safety and infrastructure on 80 km/h roads and intersections: using Accident Prediction Models," in *Young Researchers Seminar,* Torino, 2009.

[101] R. Eenink, M. Reurings, R. Elvik, J. Cardoso, S. Wichert and C. Stefan, "Accident Prediction Models and Road Safety Impact Assessment: recommendations for using these tools," European Commission, Brussels, 2007.

[102] R. F. Randolph and C. M. K. Boldt, "Safety Analysis of Surface Haulage Accidents," in *27th Annual Institute on Mining Health, Safety and Research,* Blacksburg, 1996.

[103] R. Jois, "Haul road design and road safety," in *QRC H&S Conference*, Queensland, Australia, 2011.

[104] D. Vagaja, "Safe systems approach for mining road safety," in *QRC H&S Conference*, Queensland, Australia, 2010.

[105] K. Duivenvoorden, The relationship between traffic volume and road safety on the secondary road network, Leidschendam: SWOV, 2010.

[106] United States Department of Labor, "Mine Safety and Health Administration (MSHA)," US Mines - Fatality and All-Injury Rates CY 1977 – 2015, 2017. [Online]. Available: https://www.msha.gov/.

[107] Health and Safety Executive, "Safe use of belt conveyors in mines," Crown, London, 1993.

[108] G. Homce and J. Cawley, "Electrical injuries in the US mining industry, 2000-2009," *Transactions of Society for Mining, Metallurgy and Exploration Inc.,* vol. 34, pp. 367-375, 2013.

[109] A. Asfaw, C. Mark and R. Pana-Cryan, "Profitability and occupational injuries in U.S. underground coal mines," *Accident Analysis and Prevention,* vol. 50, pp. 778-789, 2013.

[110] Safe Work Australia, "Lost time injury frequency rates (LTIFR)," 18 July 2017. [Online]. Available: https://www.safeworkaustralia.gov.au/statistics-and-research/lost-time-injury-frequency-rates-ltifr. [Accessed 08 September 2017].

[111] R. H. Peters, "Strategies for improving miners' training," NIOSH–Publications Dissemination, Cincinnati, 2002.

[112] Vensim, "Vensim® Applications," Vensim Ventana System Inc., [Online]. Available: http://vensim.com/vensim-applications/#vensim-model-reader. [Accessed 07 08 2018].

[113] T. L. Saaty , "Decision making with the analytic hierarchy process," *International Journal of Services Science,* vol. 1, pp. 83-98, 2008.

[114] M. Brunelli, Introduction to the Analytic Hierarchy Proces, SpringerBriefs in Operations Research, 2015.

[115] T. L. Saaty, Decision Making for Leaders: The Analytic Hierarchy Process for Decisions in a Complex World, Pittsburgh: RWS, 2012.

[116] E. Mu and M. Pereyra-Rojas, Practical Design Making using Super Decisions V3; An Introduction to the Analytical Hierarchy Process, Pittsburgh, PA, USA: SpringerBriefs in Operations Research, 2018, pp. 7-22.

[117] T. L. Saaty, Theory and Applications of the Analytic Network Process: Decision Making with Benefits, Opportunities, Costs, and Risks, Pittsburg: RWS, 2005.

[118] M. Alexander, "Decision-Making using the Analytic Hierarchy Process (AHP) and SAS/IML," in *SESUG*, 2012.

[119] T. L. Saaty, "Decision-making with the AHP: Why is the principal eigenvector necessary," *European Journal of Operational Research,* vol. 145, pp. 85-91, 2003.

[120] T. L. Saaty, "A scaling method for priorities in hierarchical structures," *Journal of Mathematical Psychology,* vol. 15, pp. 234-281, 1977.

[121] S. Jarek, "Removing inconsistency in pairwise comparison matrix in the AHP," *Multiple Criteria Decision Making,* vol. 11, pp. 63-76, 2016.

[122] Investing.com, "Investing.com," [Online]. Available: https://www.investing.com/commodities/copper-historical-data. [Accessed 13 08 2018].

[123] Catterpillar, 777D Off-Highway Truck, USA: Caterpillar, 2007.

[124] V. Kecojevic and D. Komljenovic, "Haul truck fuel consumption and CO2 emission under various engine load conditions," *Mininig Engineering,* pp. 44-48, 2010.

[125] M. Samavati, New Long-term Production Scheduling Methodologies for Open-pit Mines, New South Wales, Canberra, Australia: The University of New South Wales, 2017.

[126] M. Samavati, D. Essam, N. Nehring and R. Sarker, "Open-Pit Mine Production Planning and Scheduling: A Research Agenda," in *Data and Decision Sciences in Action*, Australia, Springer International Publishing AG, 2018, pp. 221-226.

[127] R. L. Lowrie, SME Minining Reference Handbook, Littleton, Colorado, USA: Society for Mining, Metallurgy, and Exploration, Inc., 2009.

[128] S. Ojha, B. K. Pal and B. B. Biswal, "OEE as an indicator for performance measurement in coal handling plant," *International Journal of Innovative Science, Engineering & Technology,* vol. 2, pp. 103-110, 2015.

[129] W. Hustrulid, M. Kuchta and R. Martin, Open Pit Mine Planning and Design, Boca Raton, Florida: Taylor & Francis Group, 2013.

[130] T. A. Williams, B. Chigoy, J. Borowiec and B. Glover, "Methodologies Used to Estimate and Forecast Vehicle Miles Traveled (VMT)," Texas A&M Transportation Institute, Texas, 2016.

[131] R. Collman, "Acoustical Control Engineers and Consultants," 25 Feb. 2015. [Online]. Available: https://www.acoustical.co.uk/acoustitips/i-sound-power-sound-pressure-technical/. [Accessed 2019 May 2].

Appendix I: Technical parameters and equations

Abbreviation	Description	Type of parameter	Equation	Unit	Reference
A_S	Shovel availability	C		%	
A_T	Truck availability	C		%	
A_{IPCC}	IPCC availability	C		%	
A_{CB}	Conveyor belt availability	C		%	
A_{Sp}	Spreader/Stacker availability	C		%	
A_i	Coefficient related to the diameter of the idler	C		-	
ARD_{IPCC}	IPCC advancement rate in depth	In	$f(TP_{IPCC}, Int_R, D_R)$	m/year	
ARD_F	Average face advancement rate in depth	In	$\dfrac{(El_{surface} - El_{bottom})}{ML}$	m/year	
ARL_{CB}	Advancement rate of conveyor belt	In	$\dfrac{ARD_{IPCC}}{\sin(I_{CB})}$	m	
B_w	Width of belt	Ax	$f(P_f, P_l, S_l, SA)$	in	[67]
C_T	Truck capacity	C		m³	
C_{IPCC}	IPCC capacity	Ax	$\dfrac{PPD}{24 \times A_{IPCC} \times U_{IPCC}}$	t/h	
C_{HBS}	Shovel bucket heaped capacity	Ax	$\dfrac{PPH_S \times t_s}{3600 \times SG_r \times C_{HBS} \times SF \times E_S \times FF \times PTF \times N_S}$	m³	[127]
c_s	Skirtboard friction factor	C		-	
D_{BP}	Distance to the base point in the surface	C		m	
D_{IPCC}	IPCC depth in the pit	L, Data	$f(ARD_{IPCC})$	m	
D_F	Face depth	L, Data	$f(ARD_F)$	m	
D_{FR}	Average distance from face to the ramp	C		m	
D_R	Depth of relocation	Ax, Data	$f(T_{IPCC})$	m	

Appendix I: Technical parameters and equations

Symbol	Description	Type	Equation	Unit	Ref
d_i	Idler diameter	C		in	
E_T	Truck operating efficiency	Ax	$A_T \times U_T$	%	[128, 127]
E_S	Shovel operating efficiency	Ax	$A_S \times A_S$	%	[128, 127]
El_{bottom}	Pit's bottom elevation	C, Data		m	
$El_{surface}$	Surface elevation	C, Data		m	
FF	Fill factor	C		%	
$FPPY$	Final product per year	Ax	$PPY \times R_{Mill} \times R_{S\&R} \times g_{ave}$	t	
g_{ave}	Average ore grade	C, Data		%	
G	Road's grade	C		%	
hp	Total conveyor power	Ax	$\dfrac{T_e \times V}{33000}$	kWh	
h_s	depth of the material touch the skirtboard	C		in	
I_{CB}	Conveyor belt inclination	C		°	
Int_R	Interval time of relocations	Ax	$f(TP_{IPCC})$	years	
K_t	Ambient temperature correction factor	C		°C	
K_x	Frictional resistance factor (between idlers and belt)	Ax	$0.00068(W_b + W_m) + \dfrac{A_i}{S_i}$	lbs/ft	[67]
K_y	Carrying run factor (combination of belt and load resistance)	C		-	
L_{CBS}	Length of each conveyor belt set	C		m	
L_{CB}	Total length of conveyor belt	Ax	$L_{CBSf} + L_{CBP}$	m	
L_{CBSf}	Length of conveyor belt at the surface	C		m	
L_{CBP}	Length of conveyor belt inside pit	L	$f(ARL_{CB})$	m	
L_b	Skirtboard length	C		ft	
ML	Mine's life	Ax	$0.2(R_M)^{0.25}$	Years	[129]
m_{UT}	Unloaded truck mass	C		t	
N_T	Number of trucks	Ax	$N_f \times \left[INT\left(\dfrac{PPD_F}{PPH_T \times WHPD}\right) + 1 \right]$	-	

Symbol	Description	Type	Equation	Unit	Ref
N_i	Number of idlers	Ax	$INT\left(\dfrac{L_{CB}}{0.0254 S_i}\right)$	-	
N_{sl}	Number of slack pulleys	C		-	
N_{tp}	Number of tight pulleys	C		-	
N_s	Number of shovels	Ax	$N_F \times N_{SF}$	-	
N_{CBS}	Number of conveyor belt sets	Ax	$\dfrac{L_{CB}}{L_{CBS}}$	-	
N_F	Number of faces	C		-	
N_{SF}	Number of shovels per face	C		-	
P_f	Percentage of fine on the conveyor belt	C		%	
P_l	Percentage of lump on the conveyor belt	C		%	
P_{LT}	Loaded truck power	Ax	$\dfrac{R_{LT} \times S_{LT}}{3.6} \cdot g$	kW	[64]
P_{UT}	Unloaded truck power	Ax	$\dfrac{R_{UT} \times S_{UT}}{3.6} \cdot g$	kW	[64]
$P_{per\,meter}$	Power consumption per meter of transferring material	Ax	$\dfrac{P_{LT} + P_{UT} + hp}{L_{CB} + TTD}$		
PPD	Production per day	Ax	$\dfrac{PPY}{WDPY}$	t	
PPY	Production per day	Ax, Data	$5(R_M)^{0.75}$	t	[129]
PPD_F	Face production per day	Ax	$\dfrac{PPD}{N_f}$	t	
PPH_T	Truck production per hour	Ax	$\dfrac{3600 \times SG_r \times E_T \times C_T}{TC_T}$	t/h	[127]
PPH_S	Shovels production per hour	Ax	C_{IPCC}	t/h	
PTF	Propel time factor	C		%	
R_{LT}	Loader truck rimpull	Ax	$(C_T \times SG_r + m_{UT})(G + R_r)$	t	[65]
R_{UT}	Unloaded truck rimpull	Ax	$m_{UT} \times R_r$	t	[65]
R_M	Mine's reserve	C		t	

Symbol	Description		Equation	Unit	Ref
R_r	Rolling resistance	C		%	
R_{Mill}	Mill recovery	C		%	
$R_{S\&R}$	Combined recovery of smelting and refining	C		%	
SA	Surcharge angle	C		°	
SF	Swell factor	C		%	
S_{LT}	Loaded truck speed	C		km/h	
S_{UT}	Unloaded truck speed	C		km/h	
SG_r	Rock density	C		t/m^3	
S_i	Idlers spacing	C		in	
S_l	Lump size on the conveyor belt	C		in	
T_e	Effective tension of conveyor belt	Ax	$LK_t\left(K_x + K_y W_b + 0.015 W_b\right) + W_m\left(LK_y \pm H\right) + T_p + T_{am} + T_{ac}$	lbs	[67]
T_{ac}	Total tensions from conveyor accessories	Ax	$T_{tr} + T_{pl} + T_{bc} + T_{sb}$	lbs	[67]
T_{am}	Tension resulting from the moving force	Ax	$\dfrac{C_{IPCC} \times 2000}{3600 \times 32.2} \times \dfrac{V - V_0}{60}$	lbs	[67]
T_p	Tension resulting from flexured belt and rotating pulleys around bearings	Ax	$200 N_{tp} + 150 N_{sl} + 100 N_i$	lbs	[67]
T_{bc}	Effective tension from belt cleaning devices	C		lbs	
T_{pl}	Effective tension from frictional resistance of plows	C		lbs	
T_{sb}	Effective tension from skirtboard friction	Ax	$L_b\left(C_s h_s{}^2 + 6\right)$	lbs	[67]
T_{tr}	Effective tension from trippers and stackers	C		lbs	
TP_m	Type of material	C		-	
TC_T	Truck cycle time	Ax	$t_D + t_{LTT} + t_{MU} + t_{MF} + t_{LT}$	sec	[127]
t_D	Delay time	C		sec	
t_{LT}	truck loading time	C		sec	

t_{LTT}	Loaded truck travel time	Ax	$3.6 \times \dfrac{TTD}{S_{LT}}$	sec	[127]
t_{MU}	Maneuver and unloading time (at crusher or dump)	C		sec	
t_{MF}	Maneuver time at the face	C		sec	
t_{UT}	Unloaded truck travel time	Ax	$3.6 \times \dfrac{TTD}{S_{UT}}$	sec	[127]
t_m	Hours of maintenance	C		h	
t_S	Shovel load cycle time	C		sec	
TP_{IPCC}	Type of IPCC	C		-	
TTD	Total truck traveling distance	Ax	$f(D_F, D_{IPCC}, D_{FR}, D_{BP}, G)$	m	
U_S	Shovel utilization	C		%	
U_T	Truck utilization	C		%	
U_{IPCC}	IPCC utilization	C		%	
U_{CB}	Conveyor belt utilization	C		%	
U_{Sp}	Spreader/Stacker utilization	C		%	
V_0	Initial conveyor speed	C		ft/min	
V	Conveyor belt speed	Ax	$f(TP_m, B_w)$	ft/min	[67]
W_b	Weight of belt	Ax	$f(SG_r, B_w)$	lbs/ft	[67]
W_m	Weight of material	Ax	$\dfrac{33.33 \times C_{IPCC}}{V}$	lbs/ft	[67]
$WHPD$	Working hour per day	Ax	$24 - t_m$	h	
$WDPY$	Working days per year	C		days	

C: constant
Ax: auxiliary
In: inflow
L: level (stock)
Data: imported data from other sources

Appendix II: Economic parameters and equations

Abbreviation	Description	Type of parameter	Equation	Unit	Reference
CC_T	Truck Capital costs	Ax	$C_{PT} + C_{RT}$	\$	
CC_{CB}	Conveyor belt capital costs	Ax	$C_{PCS} + C_{RCB} + C_{RCS}$	\$	
C_{PT}	Truck purchase cost	Ax	$f(i, p_T)$	\$	
C_{RT}	Truck replacement cost	Ax	$f(i, C_{RT})$	\$	
C_{PCS}	Conveying system purchase cost	Ax	$f(p_{CB}, L_{CB}, p_{CS})$	\$	
C_{RCB}	Conveyor belt replacement cost	Ax	$f(i, p_{CB})$	\$	
C_{RCS}	Conveyor set replacement cost	Ax	$f(i, p_{CS})$	\$	
$C_{T\&R}$	Tire and repair cost	Ax	$1.15\frac{C_t}{L_t} \times N_t \times N_T \times WDPY \times WHPD$	\$	[127]
C_{TR}	Truck repair cost	Ax	$\frac{f_R \times DV_T}{1000} \times WDPY \times WHPD$	\$	[127]
C_t	Tire cost	C		\$	
C_M	Truck maintenance cost	Ax	$0.2C_f$	\$	[127]
C_f	Fuel cost	Ax	$CNS_f \times P_f$	\$	
CNS_f	Total fuel consumption	Ax	$CNS_{fLT} + CNS_{fUT}$	lit	
CNS_{fLT}	Loaded truck fuel consumption	Ax	$0.3LF_{LT} \times P_{LT}$	lit	[127]
CNS_{fUT}	Unloaded truck fuel consumption	Ax	$0.3LF_{UT} \times P_{UT}$	lit	[127]
$C_{O\<}$	Cost of oil and lubricant in trucks	Ax	$0.1C_f$	\$	
C_{TL}	Truck labor cost	Ax	$W_{TL} \times N_t \times WDPY \times WHPD$	\$	
$C_{O\&LCB}$	Cost of oil and lubricant in conveyor belt	Ax	$0.1C_E$	\$	
C_{MCB}	Cost of conveyor belt maintenance and repair	Ax	$0.06CC_{CB}$	\$	[33]
C_{CBL}	Conveyor belt labor cost	Ax	$W_{LCB} \times N_{CBL}$	\$	
C_E	Cost of electricity	Ax	$P_{CB} \times p_e \times WHPD \times WDPY$	\$	
d	Depreciation rate	C		-	
DV_T	Depreciable value of a new truck	Ax	$f(d, p_T)$	\$	
f_R	Repair factor	C		-	

Appendix II: Economic parameters and equations

Symbol	Description	Type	Equation	Unit	Ref
i	Inflation rate	C		%	
I	Income	L	R_I	$	
L_t	Tire life	C		h	
LF_{LT}	Load factor (load truck)	C		-	
LF_{UT}	Load factor (unloaded truck)	C		-	
N_t	Number of tires per truck	C		-	
N_{CBL}	Number of conveyor belt labor force	Ax	$f(W_{LCB}, L_{CB})$	-	
OC_{CB}	Conveyor belt operating costs	Ax	$C_{MCB} + C_E + C_{O\&LCB} + C_{CBL}$	$	
OC_T	Truck operating costs	Ax	$C_{T\&R} + C_M + C_f + C_{O\<} + C_{TL} + C_{TR}$	t	
p_f	Fuel price	L	$f(i, p_{fB})$	$/L	
p_{fB}	Fuel price (base year)	C		$/L	
P_{CB}	Conveyor belt power	Ax	$0.7457 \dfrac{T_e \times V}{33000}$	kW	[67]
p_e	Electricity price	L	$f(i, p_{eB})$	$/kWh	
p_{eB}	Electricity price (base year)	C		$/kWh	
p_T	Truck price	L	$f(i, p_{TB})$	$	
p_{TB}	Truck price (base year)	C		$	
p_{ore}	Ore price	Ax		$/t	
p_{CB}	Conveyor belt price	L	$f(i, p_{CBB})$	$	
p_{CS}	Conveyor set price	L	$f(i, p_{CSB})$	$	
p_{CBB}	Conveyor belt price (base year)	C		$	
p_{CSB}	Conveyor set price (base year)	C		$	
p_{IPCC}	IPCC purchase price	Ax	$f(i, p_{IPCCB})$	$	
p_{IPCCB}	IPCC purchase price (base year)	C		$	
RC_{IPCC}	Relocation cost of IPCC	Ax	$RCM_{IPCC} \times ARD_{IPCC}$	$	
RCM_{IPCC}	Relocation cost per meter in depth	L	$f(i, RCM_{IPCC})$	$/m	
RCM_{IPCCB}	Relocation cost per meter in depth (base year)	C		$/m	
R_{OC}	Operating costs rate	In	$OC_{CB} + OC_T$	$/year	
R_I	Income rate	In	$FPPY \times p_{ore}$	$/year	
R_{CC}	Capital cost rate	In	$CC_T + CC_{CB}$	$/year	
TCC	Total capital costs	L	$f(R_{CC})$	$	

131

TOC	Total operating costs	L	$f(R_{OC})$	\$
W_{TD}	Wage of truck driver	L	$f(i, W_{TDB})$	\$/h
W_{TDB}	Wage of truck driver (base year)	C		\$/h
W_{LCB}	Wage of conveyor belt labor force	L	$f(i, W_{BLCB})$	\$/h
W_{BLCB}	Wage of conveyor belt labor force (base year)	C		\$/h

Appendix III: TEcESaS Indexes and Sustainability Index software

The software can be found and freely downloaded in the Mendeley Datasets in the following links:

TEcESaS Indexes Software: https://data.mendeley.com/datasets/b75sdckjg2/2

Sustainability Index Software: https://data.mendeley.com/datasets/kxkcmvdgw7/2

Appendix IV: Selected transportation system in the deterministic and stochastic modes

Table IV-1. Selected transportation system in the single expert (deterministic mode)

Year	Sustainability Index (SI)					Best SI	Selected System
	Truck-Shovel	FIPCC	SFIPCC	SMIPCC	FMIPCC		
2016	0.245	0.199	0.186	0.166	0.205	0.245	Truck-Shovel
2017	0.213	0.197	0.186	0.169	0.21	0.213	Truck-Shovel
2018	0.21	0.194	0.185	0.17	0.22	0.22	FMIPCC
2019	0.228	0.194	0.186	0.171	0.24	0.24	FMIPCC
2020	0.222	0.191	0.184	0.171	0.248	0.248	FMIPCC
2021	0.214	0.187	0.181	0.173	0.261	0.261	FMIPCC
2022	0.206	0.189	0.183	0.17	0.272	0.272	FMIPCC
2023	0.202	0.187	0.182	0.169	0.282	0.282	FMIPCC
2024	0.205	0.185	0.18	0.17	0.285	0.285	FMIPCC
2025	0.195	0.178	0.174	0.164	0.315	0.315	FMIPCC
2026	0.191	0.172	0.168	0.173	0.324	0.324	FMIPCC
2027	0.188	0.17	0.167	0.172	0.33	0.33	FMIPCC
2028	0.186	0.171	0.168	0.17	0.332	0.332	FMIPCC
2029	0.173	0.172	0.168	0.17	0.339	0.339	FMIPCC
2030	0.166	0.171	0.167	0.169	0.279	0.279	FMIPCC
2031	0.161	0.17	0.167	0.169	0.284	0.284	FMIPCC
2032	0.158	0.163	0.159	0.161	0.309	0.309	FMIPCC
2033	0.159	0.158	0.155	0.167	0.308	0.308	FMIPCC
2034	0.159	0.158	0.154	0.156	0.323	0.323	FMIPCC
2035	0.162	0.153	0.152	0.153	0.329	0.329	FMIPCC
2036	0.158	0.152	0.15	0.151	0.341	0.341	FMIPCC
2037	0.153	0.149	0.148	0.15	0.349	0.349	FMIPCC
2038	0.149	0.15	0.148	0.149	0.355	0.355	FMIPCC
2039	0.187	0.161	0.155	0.151	0.366	0.366	FMIPCC
2040	0.181	0.163	0.155	0.15	0.374	0.374	FMIPCC
2041	0.181	0.16	0.153	0.148	0.381	0.381	FMIPCC
2042	0.172	0.159	0.152	0.149	0.39	0.39	FMIPCC
2043	0.169	0.158	0.151	0.148	0.398	0.398	FMIPCC
2044	0.167	0.155	0.149	0.147	0.407	0.407	FMIPCC
2045	0.166	0.155	0.148	0.144	0.419	0.419	FMIPCC
2046	0.165	0.154	0.148	0.144	0.423	0.423	FMIPCC
2047	0.163	0.153	0.147	0.143	0.429	0.429	FMIPCC
2048	0.164	0.157	0.149	0.145	0.419	0.419	FMIPCC

Table IV-2. Selected transportation system in the group decision making (deterministic mode)

Year	Sustainability Index (SI)						Selected System
	Truck-Shovel	FIPCC	SFIPCC	SMIPCC	FMIPCC	Best SI	
2016	0.226	0.201	0.194	0.183	0.196	0.226	Truck-Shovel
2017	0.206	0.201	0.195	0.186	0.194	0.206	Truck-Shovel
2018	0.207	0.200	0.194	0.186	0.199	0.207	Truck-Shovel
2019	0.205	0.200	0.196	0.188	0.196	0.205	Truck-Shovel
2020	0.203	0.199	0.195	0.188	0.202	0.203	Truck-Shovel
2021	0.196	0.197	0.193	0.189	0.21	0.21	FMIPCC
2022	0.208	0.199	0.196	0.186	0.223	0.223	FMIPCC
2023	0.207	0.197	0.194	0.185	0.228	0.228	FMIPCC
2024	0.211	0.196	0.193	0.186	0.227	0.227	FMIPCC
2025	0.202	0.191	0.189	0.182	0.251	0.251	FMIPCC
2026	0.199	0.188	0.186	0.191	0.251	0.251	FMIPCC
2027	0.199	0.188	0.186	0.19	0.252	0.252	FMIPCC
2028	0.198	0.19	0.188	0.187	0.25	0.25	FMIPCC
2029	0.189	0.19	0.189	0.188	0.255	0.255	FMIPCC
2030	0.186	0.190	0.188	0.187	0.223	0.223	FMIPCC
2031	0.183	0.190	0.188	0.188	0.223	0.223	FMIPCC
2032	0.178	0.186	0.183	0.182	0.245	0.245	FMIPCC
2033	0.181	0.181	0.18	0.188	0.239	0.239	FMIPCC
2034	0.182	0.182	0.18	0.179	0.249	0.249	FMIPCC
2035	0.184	0.18	0.179	0.177	0.25	0.25	FMIPCC
2036	0.183	0.18	0.178	0.177	0.256	0.256	FMIPCC
2037	0.18	0.178	0.177	0.177	0.259	0.259	FMIPCC
2038	0.178	0.178	0.178	0.176	0.261	0.261	FMIPCC
2039	0.202	0.184	0.181	0.170	0.269	0.269	FMIPCC
2040	0.199	0.186	0.18	0.175	0.273	0.273	FMIPCC
2041	0.201	0.184	0.178	0.175	0.276	0.276	FMIPCC
2042	0.194	0.184	0.179	0.176	0.281	0.281	FMIPCC
2043	0.192	0.183	0.179	0.176	0.283	0.283	FMIPCC
2044	0.187	0.183	0.178	0.175	0.289	0.289	FMIPCC
2045	0.188	0.182	0.177	0.173	0.296	0.296	FMIPCC
2046	0.188	0.181	0.177	0.173	0.298	0.398	FMIPCC
2047	0.187	0.18	0.177	0.173	0.301	0.301	FMIPCC
2048	0.188	0.186	0.18	0.176	0.289	0.289	FMIPCC

Table IV-3. Selected transportation system in the group decision making (stochastic mode)

	Probability of Selection (PS)						
Year	Truck-Shovel	FIPCC	SFIPCC	SMIPCC	FMIPCC	Best PS	Selected System
2016	7%	4%	0%	0%	90%	90%	FMIPCC
2017	8%	4%	0%	0%	88%	88%	FMIPCC
2018	9%	0%	0%	0%	90%	90%	FMIPCC
2019	12%	0%	0%	0%	88%	88%	FMIPCC
2020	15%	0%	0%	0%	85%	85%	FMIPCC
2021	16%	0%	0%	0%	84%	84%	FMIPCC
2022	17%	0%	0%	0%	83%	83%	FMIPCC
2023	18%	0%	0%	0%	82%	82%	FMIPCC
2024	21%	0%	0%	0%	79%	79%	FMIPCC
2025	19%	0%	0%	0%	81%	81%	FMIPCC
2026	14%	0%	0%	0%	86%	86%	FMIPCC
2027	14%	0%	0%	0%	86%	86%	FMIPCC
2028	15%	0%	0%	0%	85%	85%	FMIPCC
2029	12%	0%	0%	0%	88%	88%	FMIPCC
2030	8%	3%	0%	0%	89%	89%	FMIPCC
2031	6%	6%	0%	0%	88%	88%	FMIPCC
2032	7%	5%	0%	0%	88%	88%	FMIPCC
2033	8%	3%	0%	0%	89%	89%	FMIPCC
2034	10%	1%	0%	0%	89%	89%	FMIPCC
2035	11%	0%	0%	0%	89%	89%	FMIPCC
2036	12%	0%	0%	0%	88%	88%	FMIPCC
2037	11%	0%	0%	0%	89%	89%	FMIPCC
2038	10%	1%	0%	0%	89%	89%	FMIPCC
2039	10%	0%	0%	0%	90%	90%	FMIPCC
2040	11%	0%	0%	0%	89%	89%	FMIPCC
2041	10%	0%	0%	0%	90%	90%	FMIPCC
2042	10%	0%	0%	0%	90%	90%	FMIPCC
2043	9%	0%	0%	0%	91%	91%	FMIPCC
2044	6%	3%	0%	0%	91%	91%	FMIPCC
2045	6%	1%	0%	0%	92%	92%	FMIPCC
2046	7%	1%	0%	0%	92%	92%	FMIPCC
2047	7%	2%	0%	0%	91%	91%	FMIPCC
2048	6%	3%	0%	0%	91%	91%	FMIPCC